Applied
Infrared Spectroscopy

CHEMICAL ANALYSIS

A SERIES OF MONOGRAPHS ON ANALYTICAL CHEMISTRY AND ITS APPLICATIONS

Editors

P. J. ELVING · J. D. WINEFORDNER

Editor Emeritus: I. M. KOLTHOFF

VOLUME 54

A WILEY-INTERSCIENCE PUBLICATION

JOHN WILEY & SONS

New York / Chichester / Brisbane / Toronto

Applied Infrared Spectroscopy

Fundamentals, Techniques, and Analytical Problem-solving

A. LEE SMITH

Dow Corning Corporation

A WILEY–INTERSCIENCE PUBLICATION

JOHN WILEY & SONS

New York · Chichester · Brisbane · Toronto

Library of Congress Cataloging in Publication Data:

Smith, Albert Lee, 1924–
  Applied infrared spectroscopy.

  (Chemical analysis)
  "A Wiley-Interscience publication."
  Includes bibliographical references and index.
  1. Infra-red spectrometry. I. Title. II. Series.
[DNLM: 1. Spectrophotometry, Infrared. 2. Chemistry,
Analytical. QD95 S642a]

QD96.I5858   543'.085   78-27198
ISBN 0-471-04378-8

Printed in the United States of America

10 9 8 7 6 5 4 3 2 1

PREFACE

After reaching a state of apparent maturity during the 1960s, infrared (IR) spectroscopy is now in a new period of robust growth. Sparking the revived interest is a variety of sophisticated new instrumentation that has led to experiments and analyses that would have been thought impossible only a few years ago. The potential for future accomplishments is even greater; these are indeed exciting times for IR spectroscopy.

The present volume is an expanded version of a chapter written for Kolthoff and Elving's *Treatise on Analytical Chemistry*. I think that it will be useful as a separate publication, as it fills a need not met by most other monographs. It is a volume based on practical experience, stressing the fundamental concepts and limitations of analytical IR spectroscopy for the chemist. It also provides extensive literature references for those who wish to explore more deeply a particular aspect of this subject. An alternative approach, a comprehensive coverage of all aspects of IR spectroscopy in one volume, would be encyclopedic in scope; it would also be obsolete before it was printed.

In keeping with the analytical approach, only enough theory is developed to give the reader a feeling for the basic principles of molecular dynamics. Good technique is critical if one wishes to obtain the maximum information from the spectrum; computerized instrumentation cannot compensate for casual or careless manipulation of samples or spectrometer. Therefore, considerable emphasis is given to this topic. Because it is important that the user of an IR spectrometer understand how his instrument operates, the basic design principles of existing spectrometers are described briefly. Listings of references to literature spectra and to discussions and reviews of special topics should prove useful. Since quantitative analysis by IR spectroscopy seems to be an underused and misunderstood technique, it is highlighted, and several examples are given to illustrate a variety of quantitative analyses. Factors affecting group frequencies are reviewed, but individual group frequencies are not discussed in great detail, first because excellent treatments of this topic are already available, and second because it is the author's conviction that overreliance on tables and charts, to the exclusion of the study of reference spectra, is a mistake. Because this book is analytically oriented, not discussed are a number of interesting topics that seem to fall more in the realm of physical chemistry—although the dividing line between analytical and physical chemistry is admittedly a hazy one. Not included are such topics as chemisorption and the mechanism of heterogeneous catalysis, studies of molecular structure, rotation–vibration spectra analysis, and the determination of energy-band separation in semiconductors.

Throughout the book fundamental concepts and limitations are stressed, so that the reader is provided with a sound basis for pursuing his own more specialized interests.

It is customary and appropriate to thank those who have contributed to such a work as this. I have had the good fortune to know and to have worked with some of the best applied spectroscopists in the world. To them I owe a great deal, for their instruction, advice, and encouragement. I would particularly like to acknowledge the help of Clara Craver, Peter Griffiths, N. J. Harrick, R. W. Hannah, and R. A. Nyquist, who have read portions of the manuscript and offered helpful comments. Finally, I would like to thank Miss Gertrude Binder and Mrs. Myrna Freeman for their dedicated effort in typing the manuscript.

A. LEE SMITH

Midland, Michigan
January 1979

CONTENTS

Applied
Infrared Spectroscopy

INTRODUCTION

SCOPE OF IR

Infrared spectroscopy (IR) is used by workers in many disciplines, but the term carries a different meaning in each field. To the analytical chemist, it is a convenient tool for solving problems such as determination of the five isomers of hexachlorobenzene; characterization of the wax, resin, polymer, and emulsifier in emulsion polishes; and identification of the country of origin of illegal opium. To the physicist, it may represent a method for studying energy levels in semiconductors, determining interatomic distances in molecules, and measuring the temperature of rocket flames. To the organic chemist, it furnishes a way of fingerprinting organic compounds, picking out functional groups in an organic molecule, and following the progress of a reaction. To the biologist, it promises a method of studying transport of bioactive materials in living tissue, provides a key to the structure of many natural antibiotics, and gives many clues in the study of cell structure. To the physical chemist, it can furnish a revealing look at mechanism of heterogeneous catalysis, provide a convenient means for following the kinetics of complicated reactions, and serve as an aid to determining crystal structures. In these fields and many others, IR spectroscopy provides researchers with powerful insights. It is probably fair to say that IR spectroscopy is the most nearly universally useful of all instrumental techniques.

HISTORY

The existence of IR radiation was first recognized in scientific literature in 1800 by Sir William Herschel [3], who used a glass prism with blackened thermometers as detectors to measure the heating effect of sunlight within and beyond the boundaries of the visible spectrum. Although Herschel first concluded that IR radiation was of the same nature as visible light, he later reversed his belief because of the two different curves he obtained for "luminosity" and heating effect. Herschel's main interest was astronomy, however, and he did not further investigate the "heat spectrum."

Interest in this phenomenon lay dormant for the next 80 years, but during 1882-1900 several investigators, using the comparatively crude methods available to them, made brief forays into the IR region. Abney and Festing [1], for example, photographed absorption spectra for 52 compounds to 1.2 μm and correlated absorption bands with the presence of certain organic groups in the molecule. Julius [5] investigated the spectra of 20 organic compounds using a rock-salt (NaCl) prism and bolometer detector. He found that compounds containing a methyl group absorbed at 3.45 μm, but his measurements at longer wavelengths were grossly in error.

The theory of IR absorption had not yet been developed, and up to that time it was by no means settled whether the absorption was due to the presence of individual atoms within the molecules, to intermolecular effects, or to intramolecular motions. Julius' work indicated that the latter explanation held, specifically, that the grouping of atoms in a chemical composition determines the absorption pattern.

Activity in the field was growing, but it remained for W. W. Coblentz to lay the real groundwork for IR spectroscopy. During his classic researches starting about 1903, he investigated the IR spectra of hundreds of substances, both organic and inorganic [2]. His work covered the rock-salt region with a thoroughness and accuracy such that many of his spectra are still usable today.

The experimental difficulties of the early researchers were enormous. They not only had to design and construct their own instruments but all the components as well. This included grinding and polishing the prism, silvering the mirrors, and constructing the radiometer. The instruments were calibrated by using values for the indices of the refraction of rock salt measured by themselves or other investigators. Spectrometers were usually located on the basement floor and measurements often made at night, to minimize the effect of vibrations on the sensitive radiometer or radiomicrometer. Since each point in the spectrum had to be measured separately and at least 10 points per micrometer were measured, obtaining a spectrum was a tedious job requiring 3-4 h or more [4].

Out of this early work came the recognition that each compound had its own unique IR absorption pattern and that certain groups, even in different molecules, gave absorption bands that were found at approximately the same wavelength. Because of the difficulty of measuring the spectra, however, the technique was virtually unused by chemists until the 1940s.

World War II brought not only an expanded need for analytical instrumentation, but also a rapid advance in the science of electronics. It became possible to amplify electronically the very small signals obtained from a tiny thermocouple in an IR spectrometer and to record them on a strip chart. Because the thermocouple was slow in response, a direct-current (DC) signal was used and drift was a serious problem. Nevertheless, with luck and care, spectra of quite good quality could be obtained in a matter of 1–2 h.

The next significant improvement came with the technique of making thermocouple detectors with response rapid enough for them to be used with radiation that was chopped five to ten times a second. This advance made possible elimination of drift in the spectrometer record and opened the door to the possibility of double-beam spectrometers, which could be programmed to give charts reading in percent transmittance as a function of linear wavelength or wavenumber [7].

This development took IR spectroscopy out of the realm of tedium, and as chemists became aware of its potential, popularity of the technique expanded rapidly. After a period of intensive effort during which thousands of materials were examined and instrumental limitations explored, spectroscopists began to demand better resolution and wavelength accuracy. Grating dispersion answered at least some of these demands, and with the advent of interferometer spectrometers, the possibility of obtaining accurate band contours in liquids and solids is now in sight. Perhaps it is not too much to hope that it will soon be possible to obtain noise-free spectra that truly represent the contours, frequencies, and intensities of the molecular absorption pattern, undistorted by spectrometer or sample cell artifacts.

APPLICABILITY

The IR absorption spectrum of a compound is probably its most unique physical property. Except for optical isomers, no two compounds having different structures have the same IR spectra. In some cases, such as with polymers differing only slightly in molecular weight, the differences may be virtually indistinguishable but, nevertheless, they are there. In most instances the IR spectrum is a unique molecular fingerprint that is easily distinguished from the absorption patterns of other molecules.

In addition to the characteristic nature of the absorption, the magnitude of the absorption due to a given species is directly relatable to the concentration of that species. Thus measurement of absorption intensity gives, on simple calculation, the amount of a given constituent present in the sample.

The technique is almost universal in scope. Samples can be liquids, solids, or gases. They can be organic or inorganic, although inorganic materials sometimes do not give very definitive spectra. The only mole-

cules transparent to IR radiation under ordinary conditions are mona-
tomic and homopolar molecules such as Ne, He, O_2, N_2, and H_2.

Another limitation is that the ubiquitous solvent water is a very strong
absorber and, furthermore, attacks ordinary rock-salt sample cells. Infrared is usually not very sensitive to impurities in a sample at levels
below 1%. This can be a blessing or a curse, depending on one's viewpoint and the problem at hand. Similarly, the fact that positions of the
characteristic absorption maxima for different groups are not quite constant from one molecule to another may be frustrating, but it accounts
for the uniqueness of the absorption pattern and furnishes more clues to
molecular structure than if the bands were invariant.

Spectroscopy in the IR region has some special difficulties. Because
the radiation is invisible, optical materials cannot be evaluated by the
eye. The energies involved are extremely low and become lower at longer
wavelengths.

As a consequence of this low energy, the signal from the detector is
not very much greater than the noise arising from random thermal motions of electrons within the detecting circuit. Furthermore, since all the
components of the spectrometer are warm (compared to the absolute
zero) and thus radiate IR energy, large amounts of false radiation reach
the detector and must be distinguished from the true signal. Someone has
said that IR spectrometry is roughly comparable to photographing an
emission spectrum using a white-hot spectrograph. In view of these
difficulties, Dr. Samuel Johnson's comment comparing women preaching
to dogs walking on their hind legs might aptly apply to the obtaining of
spectra by the early IR pioneers; it was not done very well, but one is
surprised to find it done at all.

SELECTION OF AN ANALYTICAL TECHNIQUE

As a general rule, IR spectroscopy should be used wherever specificity
is desired. A melting point, refractive index, or specific gravity gives
only a single point of comparison with other substances. By contrast, an
IR spectrum gives an almost infinite number of such points. It usually
furnishes a more definitive test for functional groups than does chemical
analysis.

Infrared is usually to be preferred when a combined qualitative and
quantitative analysis is required. Applications of this type include following organic reactions, particularly when the course of the reaction is not
well known, and characterizing the end products of such reactions.

The IR approach is indicated in quantitative analysis of complex nonvolatile mixtures such as polymers or in the cases where separation of components of a mixture is difficult using gas chromatography (GC).

On the other hand, the IR spectrum will not show differences in composition or structure that are reflected in the spectrum as variations of the same magnitude as random instrumental errors. It will not indicate whether a material will meet performance specifications, except as performance can be related to the presence or absence of certain absorption bands. As with any complex data, the spectrum is capable of being misinterpreted by persons who have only a superficial knowledge of the field.

Other techniques may be faster or more specific in certain areas. For example, nuclear magnetic resonance (NMR) can often furnish more information about certain kinds of soluble organic molecules without reference spectra or standards. Standards are less important in mass spectroscopy, and smaller samples may be used, but the material must be volatile and the range of applicability is somewhat less than for IR. Gas chromatography, mass spectroscopy, and ultraviolet (UV) all give excellent sensitivity for trace analysis within their limitations. Again, for certain materials, these three techniques are capable of giving excellent quantitative results. Raman spectroscopy may be used for analytical purposes in a manner similar to IR but is more often complementary than competitive [6]. Thus the implication is clear that the analyst should recognize the capabilities and limitations of all available techniques.

DEFINITIONS

Each branch of science has its own language, and it may be helpful here to define some of the terms that are used in the present work.

A spectrum may be regarded as an arrangement of electromagnetic radiation ordered according to wavelength. These wavelengths may range from 10^{-12} mm to millions of meters. For convenience, the terms angstrom unit (Å) = 10^{-8} cm and micrometer (μm), formerly micron (μ), = 10^{-4} cm are used. A term often encountered in IR spectroscopy is wavenumber (ν), whose relationship to the wavelength λ is $\nu(\text{cm}^{-1}) = (10^4/\lambda)$ (micrometers). The wavenumber may be visualized as the number of integral wavelengths of electromagnetic energy per centimeter (Fig. 1.1). Wavenumber is directly proportional to energy and to the frequency of the vibrating unit emitting the radiation:

$$\sigma = \frac{f}{c} = \frac{E}{hc} = \frac{1}{\lambda_{\text{vac}}} = \frac{\nu}{n} \qquad (1.1)$$

where σ = vacuum wavenumber in cm^{-1}; f = frequency in seconds

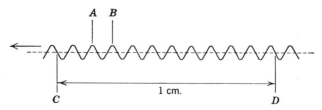

Fig. 1.1 Relationships of units. Wavelength corresponds to distance AB; wavenumber is number of waves per centimeter, CD; and frequency is the number of waves passing fixed point C in unit time.

(physicists often use $\bar{\nu}$ to represent frequency, but we use f to avoid confusion with the wave-number symbol ν); E = energy in ergs; h = Planck's constant; λ_{vac} = wavelength in centimeters measured in (or corrected to) vacuum; and n is the refractive index of air. From a fundamental standpoint it is more significant to speak in terms of frequency or vacuum wavenumber, since the wavelength of radiation depends on the refractive index of the medium in which it is measured. Frequency, on the other hand, is independent of the medium through which the radiation is passing.

The IR region of the spectrum is usually considered to start near the red end of the visible spectrum at the point where the eye no longer responds to dispersed radiation (thus *infra*, or below, the red), which is about 14,000 cm^{-1} (7000 Å or 0.7 μm). The so-called fundamental IR region begins at about 3600 cm^{-1} or 2.8 μm. The analytically useful IR region extends from 3600 cm^{-1} to somewhere around 300 cm^{-1} or 33 μm. This book is primarily concerned with the mid-IR, although the "near" IR region between 14,000 cm^{-1} and 3600 cm^{-1} is also considered briefly. The "far" IR region is not easily defined but approximately covers the 300 cm^{-1} to 20 cm^{-1} span. It is not used much for analytical purposes. Radiation at frequencies lower than 20 cm^{-1} (500 μm) is classified as being in the microwave or radiowave region.

References

1. Abney, W. deW., and E. R. Festing, *Phil. Transact.*, **172**, 887 (1882).

2. Coblentz, W. W., *Investigations of Infrared Spectra,* Carnegie Institute, Washington, D.C., Publication No. 35, 1905 (reprinted 1962 by Coblentz Society and Perkin Elmer Corporation).

3. Herschel, W., *Phil. Transact. Roy. Soc., * **90**, 284 (1800).

4. Jones, R. N., *Appl. Opt., * **2**, 1090 (1963).

5. Julius, W. H., *Verhandl. Konikl. Acad. Wetenschappen Amsterdam,* **I,** 1 (1892).

6. Sloane, H. J., *Appl. Spectrosc.,* **25,** 430 (1971).

7. Wright, N., and L. W. Hersher, *J. Opt. Soc. Am.,* **37,** 211 (1947).

INSTRUMENTATION

The proliferation of spectrometer types presents a confusing picture even to the experienced spectroscopist. Yet the principles of the various designs are not difficult to understand. We briefly discuss the currently available systems so that the reader can pursue more detailed accounts, if he wishes, with a minimum of confusion. It is useful first to adapt Winefordner's detector classification scheme [86] to spectrometer systems, as shown in Fig. 2.1. *Sequential devices* are those in which information is collected sequentially in time. They utilize a single detector to collect information as each spectral element is scanned in turn. *Spatial devices,* which utilize multiple detectors, are not common in the mid-IR; in the visible region, a photographic recording instrument (spectrograph) is an example of spacial recording device. *Multiplex devices* are those in which a single viewer (detector) receives many signals simultaneously from different elements of the spectrum. These signals are transmitted by a single channel but can be decoded in such a way as to give information about each of the individual spectral elements.

INFRARED PHYSICS

To understand the operation of the various types of IR spectrometers, we review some of the physical principles relevant to spectrometer design. A thorough treatment of the design, construction, and operation of IR spectrometers from a physicist's viewpoint has been provided by Stewart [79]. As we noted earlier, the IR region is intrinsically low in energy (because energy is inversely proportional to wavelength). The design of IR spectrometers is thus always biased toward maximizing energy throughput. Similarly, in the operation of a spectrometer one must always recognize this fundamental limitation and select operating parameters consistent with the demands of the laws of physics.

All IR spectrometers, whether simple or sophisticated, have certain elements in common. They are the source, optical system, detector, and amplifier.

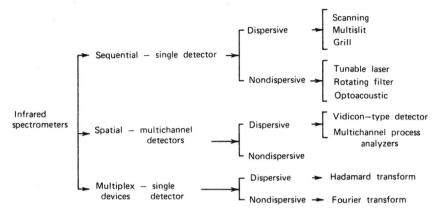

Fig. 2.1 Classification of IR spectrometers.

Source

The ideal source for IR spectroscopy would be a monochromatic emitter of high intensity, continuously tunable over a wide range of wavelengths. Although tunable lasers provide an approximation to this ideal (see p. 22), their use in general-purpose spectrometers seems somewhat remote as of this writing; sources that radiate a continuous spectrum approximating a blackbody are commonly used. The power W radiated by a blackbody is given in terms of wavelength λ and its temperature T by Planck's law:

$$W = \frac{c_1}{\lambda^5} \left[\exp\left(\frac{c_2}{\lambda T}\right) - 1 \right]^{-1} \qquad (2.1)$$

where c_1 and c_2 are constants.

In the region 100–4000 cm^{-1} the most popular sources are the Globar, which is a bonded silicon carbide rod, and the Nernst glower, which is a mixture of zirconium, thorium, and yttrium oxides. A source of slightly lower emissivity but much longer life can be made from a tightly coiled helix of Nichrome resistance wire. A plot of radiant power as a function of wavelength and temperature (Fig. 2.2) shows that raising the temperature of the source gives a large increase in radiation at short wavelengths but only a slight gain at longer wavelengths. Since increased short-wave energy means more stray light, there is usually not much advantage in going to high source temperatures. For far-IR work (<100 cm^{-1}) the high-pressure mercury lamp gives better emission than do heated filaments or rods. In any spectrometer, source dimensions should be such that the

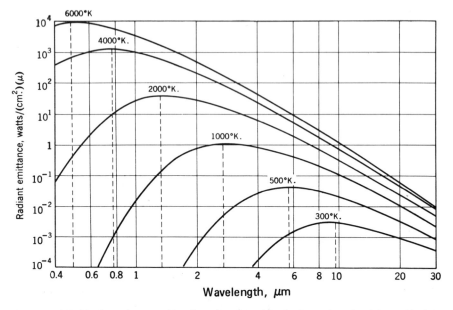

Fig. 2.2 Blackbody emittance plotted against wavelength. Reproduced, with permission of McGraw-Hill Book Company, from Hackforth [39].

source image fills the aperture of the spectrometer at all wavelengths. Infrared sources have been reviewed by Cann [7].

Optical System

The purpose of the optical system is to channel the radiation along the proper path. Front-surfaced mirrors rather than lenses are used because lenses are subject to chromatic aberration and a refracting optical system would have to be constantly readjusted with changes in wavelength.

Reflecting optics can be ground to have either flat or spherical surfaces or more complicated figures such as toroids, paraboloids, or ellipsoids (Fig. 2.3). Large-aperture optical systems usually require aspherical optics to minimize aberrations and energy losses. In particular, the collimating mirror is usually an off-axis paraboloid and the thermocouple receiver mirror is an ellipsoid. Although such shapes produce significantly better image fidelity than do spherical optics, they do not give optical perfection, even theoretically, except for a point source.

As a general rule, the larger the aperture of an optical system (i.e., the

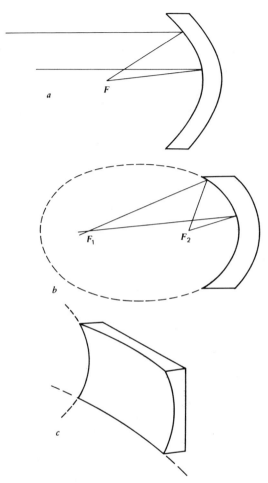

Fig. 2.3 Parabolic mirror (*a*), which focuses parallel light at *F*; elliptical mirror (*b*), with two foci, F_1 and F_2; and toroidal mirror (*c*), in which focal length differs in the horizontal and vertical planes.

shorter the optical path relative to the size of the components), the greater the demands on the quality and the adjustment of the mirrors, slits, and other elements. A large-aperture system is desirable, since more energy is then available for a given spectral slit width than with a low-aperture system. For detailed consideration of spectrometer design, the reader should consult the works by Conn [17] and Stewart [79].

Detectors

Infrared radiation detectors fall into two classes, thermal detectors and quantum or photon detectors. The former category includes the thermocouple [a heat-sensitive junction of dissimilar metals, which develops an electromotive force (EMF) that changes with temperature] and the bolometer (a resistor with a large temperature coefficient of resistance), as well as pneumatic types. The second category includes the intrinsic and impurity-activated photon detectors. The energy available in the IR region is too low to activate photographic emulsions or photoelectric detectors.

Since the thermocouple is basically a heat engine, it is grossly inefficient (efficiency $<10^{-6}$) for very small temperature changes it must detect in an IR spectrometer. The response of a thermocouple to a changing signal is slower than that of a photon detector. Thermocouples, however, do have a constant response independent of wavelength, except at longer wavelengths (beyond 30 μm), where the efficiency of the blackening decreases. For the far IR, the pneumatic detector is more efficient.

The absolute sensitivities of thermocouples and bolometers to a pulsating signal have been discussed by Williams [84]. The sensitivity of a thermocouple is roughly proportional to the inverse of the receiver area and heat capacity. Other factors affecting sensitivity include the thermoelectric coefficient and the total heat dissipation. Bolometers likewise show greater sensitivity with smaller receiver area and heat capacity. It is thus advantageous to have the receiving element as small as possible in both area and heat capacity. Further, an element with low heat capacity reacts more quickly to an intermittent or chopped signal, and as we see later, a rapid detector response is desirable. Bolometers and thermocouples are usually operated in a vacuum to increase sensitivity and decrease noise. Otherwise similar thermocouples may vary widely in sensitivity because of different thermoelectric coefficients. It is advantageous to use the most sensitive thermocouple obtainable, since energy seldom can be bought so cheaply. The radiant power falling on the detector of a spectrometer is on the order of 10^{-7} W and induces a temperature change in the element of a few thousandths of a degree. The resulting signal is a fraction of a microvolt.

The Golay or pneumatic detector [31] utilizes the thermal expansion of a nonabsorbing gas contained in a blackened receiver chamber to deform a flexible mirror. Motion of the mirror is amplified by an optical lever and is detected by a phototube (Fig. 2.4). Sensitivity is about the same order as that of the thermoelectric detectors, but focusing on the receiver

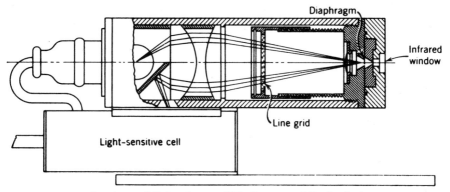

Fig. 2.4 One form of Golay detector [31]. Courtesy of Eppley Laboratory.

is not so critical and the blackening can be rendered more responsive to longer-wavelength radiation.

In photon detectors individual quanta of radiation excite electrons. In a conventional photoelectric cell these excited electrons are actually ejected from the surface, but in the IR region the photon energy is too low to cause a photocurrent in this manner. Instead, semiconductor devices are used in which electrons are raised by absorption of a photon from a nonconducting energy state or valence band into the metastable conducting state. The energy of the photons must be sufficient to "kick" the electrons across the energy gap E_g to make the detector conduct [i.e., $\nu > (E_g/h)$]. The height of the energy barrier E_g is temperature sensitive and decreases when the temperature is lowered (Fig. 2.5).

Intrinsic photon detectors such as PbS, PbSe, and InSb are useful for special applications because they are 10–100 times more sensitive than thermoelectric detectors but are effective only over limited wavelength ranges (Fig. 2.5). Their sensitivity and range can be improved by cooling them to liquid N_2 temperature.

Semiconductor (impurity-activated or extrinsic) detectors may be used down to 250 cm^{-1}, but they must be cooled to liquid N_2, or in some cases liquid helium, temperatures to take advantage of their unique properties. The extremely rapid response of these detectors makes them attractive for some applications in spite of their disadvantages.

Pyroelectric bolometer detectors such as triglycine sulfate (TGS) are used in interferometer spectrometers because of their good response to a wide range of IR frequencies and radiation intensities. They do, however, lose some efficiency at fast chopping rates. Such detectors are

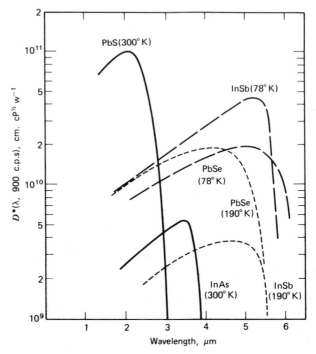

Fig. 2.5 Spectral response of intrinsic detectors. Reproduced, with permission, from *Infrared Physics* [4].

ferroelectric materials and show a strongly temperature-dependent electric polarization below their Curie temperature. This characteristic can be exploited to detect the very small thermal changes induced by radiation transmitted through the spectrometer [19]. More detailed reviews of IR detectors are given elsewhere [57, 63, 70, 78].

Amplification

Early IR spectrometers used optical amplification of the detector signal obtained from a galvanometer or radiometer. Current instrumentation uses chopped radiation with electronic amplification almost exclusively.

There are several reasons for chopping the radiation. Alternating-current (AC) amplifiers are more stable than DC amplifiers, and by tuning the amplifier to pass the chopped frequency and reject all other frequencies, one can reduce the noise level considerably. Furthermore, drift can be minimized since the alternating signal has a built-in zero reference (single-beam spectrometers) or radiation reference (double-beam spectrometers). The disadvantage is that at least half of the light is discarded, but this loss is more than compensated for by the advantages listed.

SEQUENTIAL IR SPECTROMETERS

The principles of some of the more common optical arrangements used in IR spectrometers are discussed briefly. For more detailed explanations of specific instruments, the reader is referred to the manufacturer's literature. As near-IR spectrometers are more like spectrometers for the visible region, they are not discussed in this chapter. Arrangements shown here do not necessarily represent any actual existing instrument, but most commercial spectrometers are constructed in a manner similar to one of the diagrams.

Dispersive

We consider *source optics* and *monochromator optics* separately. Source optics for a simple single-beam spectrometer are diagrammed in Fig. 2.6a. They consist of an IR source (S), a chopper (C), and a mirror (SM), to focus an image of the source on the slit. The sample is usually placed at the position where the beam area is a minimum. Chopping at the source has the advantage that true spectra can be obtained from hot or cold samples, since radiation from the sample is unchopped and hence ignored by the spectrometer.

A typical source arrangement for the double-beam optical null spectrometer is shown in Fig. 2.6b. Here both beams "see" essentially (but not exactly) the same portion of the IR heater, so the effects of its temperature fluctuations are minimized. After passing through the sample and reference cells the beams are combined at a mirror (CM), which is often a rotating semicircular mirror. An optical wedge or comb (A) is moved in and out of the reference beam by the servomechanism in such a way as to match the absorption in the sample beam. The motion of this attenuator is transmitted to the pen, which then records directly in percent transmittance.

Source optics for another type of double-beam spectrometer, using ratio recording instead of an optical attenuator, are shown in Fig. 2.6c. With this arrangement both beams are chopped, but at different frequencies, so the instrument can decode the optical signal and compare the intensities of the two beams electrically.

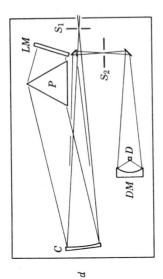
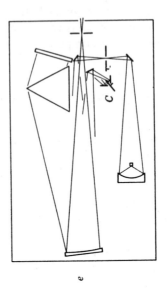
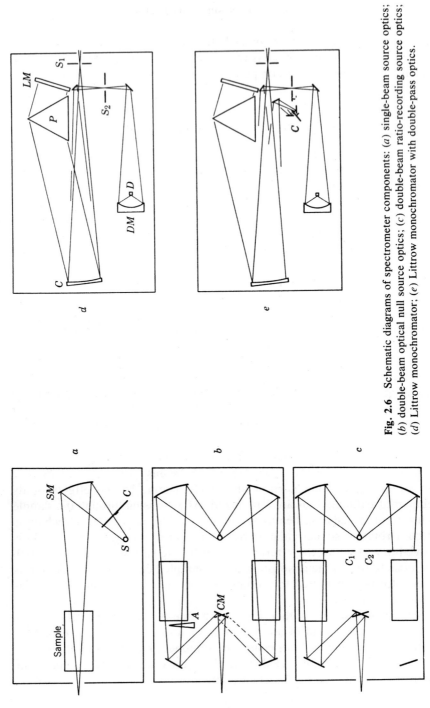

Fig. 2.6 Schematic diagrams of spectrometer components: (*a*) single-beam source optics; (*b*) double-beam optical null source optics; (*c*) double-beam ratio-recording source optics; (*d*) Littrow monochromator; (*e*) Littrow monochromator with double-pass optics.

Conventional Monochromators

The simple Littrow monochromator shown in Fig. 2.6d may be combined with any of the source optical systems to give single- or double-beam spectrometers. The light beam enters at the slit (S_1) and is collimated by the mirror (C). The prism (P) is set at the angle of minimum deviation, and the Littrow mirror (LM) is rotated by a screw or cam drive to scan the desired wavelength range. In grating spectrometers, the grating is used at position LM along with appropriate filters at S_1. The light beam retraverses the prism and is brought to a focus along the plane of S_2. At this point a spectrum is spread out along the S_2 plane. A bundle of closely related wavelengths pass through the exit slit S_2 and is focused by the mirror DM on the detector (D). The signal from D is suitably amplified and filtered and is used to drive the reference beam attenuator (optical null system) or the recorder pen (single-beam spectrometer or ratio recording system).

Vertical slits are used at both the entrance and exit of the monochromator. Their width is varied during the scan to compensate for the large change in energy with wavelength. The entrance and exit slits are usually kept at the same width by a direct coupling mechanism.

The reason for having two slits is as follows. The entrance slit limits the horizontal angle through which radiation enters the monochromator. It acts as a line source in a manner analogous to the slit of an emission spectrograph. It is well known that, for a stigmatic spectrograph, the spectral lines have the same shape as the slit. Narrowing the slit produces narrower lines (down to the Rayleigh diffraction limit).

The IR monochromator, instead of having a photographic plate as a detector, uses an exit slit that, in effect, scans across the spectrum. Here again, it is easy to see that the energy escaping to the detector will be a bundle of adjacent wavelengths that will approach monochromatic radiation as the slit width is decreased. The width of this energy band at half peak intensity is known as the *spectral slit width*.

If the monochromator had one slit, the energy passed by it would be proportional to the slit width. Since there are two slits and each passes a certain fraction of the light, the total radiation emerging is proportional to the *product* of the two slit widths. If both slits are of equal width, the energy passed by the monochromator is proportional to the *square* of the slit width. This relationship is important to keep in mind during operation of the spectrometer.

In double-beam instruments and some single-beam ones, the slits are programmed by a mechanical or electrical cam to keep the energy arriving at the detector approximately constant as the wavelength is varied. The maximum usable slit width is limited by the point at which the slit image fills the detector target.

The radiation is dispersed by either a prism or a grating. Historically, the prism has been the nucleus around which IR spectroscopy developed. Prisms made from naturally occurring minerals were initially used, but such materials are limited in size and optical quality. The development of techniques for growing large single crystals of alkali halide and other materials made possible the general use of good-quality prisms at a reasonable cost. By far the most popular prism material is rock salt, because it covers a large part of the fundamental or "fingerprint" region of IR at a reasonable compromise of cost, range, and resolution.

For a prism to disperse radiation and form a spectrum, the refractive index of the material must change with wavelength. The more rapid the change in refractive index, the better the dispersion. Since this condition holds near an absorption band, the best dispersion of a prism is obtained just before its absorption cutoff.

Prism monochromators have the advantage of being more simple than those using gratings. Disadvantages are that prism resolution is limited and varies with wavelength and that the dispersion is temperature sensitive. Prisms also introduce a wavelength-dependent distortion of the slit image; that is, the image of a straight entrance slit appears at the exit slit position as a segment of a parabola. One (or both) of the slits is usually curved to compensate for this effect, which can be the principal factor limiting the practical resolution of the spectrometer [79].

Grating instruments are inherently superior to prism spectrometers because of the improved resolution attainable, and higher resolution implies greater qualitative and quantitative accuracy. The increased resolution is a consequence of greater energy transmission; for otherwise equivalent monochromators and under optimum conditions, the luminosity or light flux passed by a grating is calculated to be 6.7 times that of a NaCl prism at 625 cm^{-1}, 25 times at 1250 cm^{-1}, 50 times at 2500 cm^{-1}, and 100 times at 5000 cm^{-1} [47]. Somewhat less than this will be realized in practice [11]. Another advantage of gratings is that resolving power does not change appreciably with wavelength.

It should be pointed out that gratings characteristically polarize the IR beams, with the amount of polarization depending on the direction of the incident beam relative to the grating surface [26, 87]. Polarization also occurs in prism spectrometers, but to a lesser degree.

Disadvantages of gratings include the necessity of eliminating unwanted spectral orders, more scattered light, less efficient use of the radiation, and a limited wavelength range for a single grating. The first two shortcomings may be overcome by use of a foreprism in a double monochromator arrangement or by the use of band-pass filters that have very low transmission at unwanted frequencies and good transmission in the region of interest. Usually several filters and two or more gratings are used to cover a complete spectrum. The band-reject requirements on the filter are quite severe, since the shorter-wavelength radiation, which accounts for the largest part of the stray light, is much more intense (see Fig. 2.2). The efficiency of a grating may be increased by cutting the grooves in such a way as to increase the intensity in a certain order (blazing).

The double monochromator used in some spectrometers is, as the name implies, two monochromators in series. The advantages to this arrangement include very low stray radiation and an increase in dispersion by a factor of almost 2. Disadvantages are the increased complexity and cost of using two monochromators instead of one.

A double-pass system, which essentially converts a single monochromator to a double monochromator, is shown in Fig. 2.6e. It is ordinarily used in the single beam mode with a simple unchopped source, since chopping at C is used to discriminate between single- and double-pass radiation, but has been combined with double-beam source optics also. The double-pass monochromator has the same dispersion and stray energy advantages as the double monochromator; but since chopping occurs *after* the beam passes through the sample, the instrument is sensitive to sample temperature. In quantitative analyses, the zero setting may shift during a scan because of heating of the sample or the slit jaws by the radiation beam.

Other monochromator arrangements are possible but are less commonly used than the Littrow system described. Some double-beam systems that can be used for IR spectrophotometry have been compared by Golay [32].

Multislit Spectrometers

First proposed in 1949 by Golay [30], multislit spectrometers have never reached the marketplace. The rationale for the multislit arrangement depends on the fact that the radiant power transmitted by the spectrometer increases in direct proportion to the number of slits. Thus, in principle, many narrow slits, giving high resolution, can be combined so as to improve the energy transmitted by the monochromator.

Operation of the multislit spectrometer may be understood by visualizing a conventional monochromator, in which a small bundle of adjacent frequencies passes through the exit slit and falls on the detector. For simplicity, we say that the radiation is nominally monochromatic and of frequency ν_0. Because the monochromator is stigmatic at ν_0, radiation passing the entrance slit at a given point will fall at a corresponding point along the exit slit. For example, if the bottom of the entrance slit is blocked, the bottom of the exit slit will be darkened. Let us now imagine a second entrance slit, also illuminated by the source, in the plane of the first slit but laterally displaced from it by a small distance d. A bundle of radiation from the second entrance slit, also of frequency ν_0, will fall on a second exit slit displaced to distance d from exit slit 1. We have now doubled the energy reaching the detector with no loss in resolution. We have also introduced a problem: radiation of some other frequency (not ν_0) is also passing through entrance slit 1 and exit slit 2, and through entrance slit 2 and exit slit 1. What is needed is a way of coding the radiation from each slit so that only radiation from corresponding entrance and exit slits is recognized by the spectrometer. Golay solved this problem by designing a system of slits cut in rotating disks that acted both as slit and chopper. The chopper was constructed in a manner such that the chopping frequency for unwanted radiation (e.g., that transmitted by entrance slit 1 and exit slit 3) was different from that for wanted radiation (e.g., entrance slit 2 and exit slit 2) and could be differentiated by a tuned amplifier. The spectrum was scanned in the usual way, namely, by rotation of the Littrow mirror of the monochromator. Golay demonstrated a 10-fold increase in throughput as against a theoretical gain of 32 with a 64-slit instrument. The gain was less than expected because of partial blockage of the beam by the chopper shaft and other mechanical losses of radiation.

Grill Spectrometers

In an extension of Golay's work, Girard [27–29] modified the openings in the entrance and exit apertures to form grills arranged in a regular pattern. One grill design is shown in Fig. 2.7. Here again, the purpose is to increase the energy throughput without decreasing the resolution. An entrance grill and an exit grill are used in place of slits. The grills transmit 50% of the incident radiation, and one grill is vibrated (e.g., displaced vertically) so that only corresponding elements of the entrance and exit radiation are chopped. For properly designed grills, luminosity gains of over 100 have been claimed.

A number of problems have been noted with grill spectrometers. First, the entrance and exit grills must be accurately but not necessarily identically fabricated; account must be taken of optical distortions in the spectrometer. Second, some sampling accessories (e.g., multipass gas cells or microcells) may cause problems, because if the radiation beam

Fig. 2.7   Hyperbolic Girard grill [28, 29].

is not uniform, large side lobes in the apparatus function may result (loss of spectral purity). Third, since a large amount of unmodulated radiation reaches the detector, any vibration, especially one of the same frequency as the vibrating grill, results in noise. The theory, advantages, uses, and problems of grill spectrometers have been discussed by Moret-Bailly [62].

Nondispersive

Dielectric Filter Spectrometers

Sophisticated vacuum coating techniques, which utilize alternating layers of dielectric materials having different refractive indices deposited on a transparent substrate, have made possible band-pass filters, cutoff filters, and spike filters (Fig. 2.8). Because the filter is composed of dielectric materials, it does not absorb the incident radiation; it only reflects and transmits. Spike filters can be constructed to have quite narrow band widths and excellent rejection of unwanted frequencies. A clever modification of the spike filter, in which the filter is constructed such that the transmitted frequency varies continuously and uniformly on a disk-shaped substrate, can be used as a monochromator. The spectrum is scanned by slowly rotating the disk. No prism or grating is required. A compact, relatively simple IR spectrometer is thus possible (Fig. 2.9).

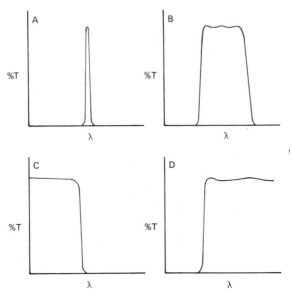

Fig. 2.8 Typical transmission spectra of different types of interference filters: (*a*) spike filter; (*b*) band-pass filter; (*c*) short-wavelength pass filter; (*d*) long-wavelength pass filter. Reprinted from Stewart [79], p. 173, by courtesy of Marcel Dekker, Inc.

Such instruments can even be made portable for use in the field. A microcomputer-controlled spectrometer designed primarily for quantitative analysis utilizes a dielectric filter monochromator [81].

Tunable Lasers

The narrow band width (ranging from 1 cm^{-1} to 10^{-6} cm^{-1}) and concentrated power of tunable lasers are irresistable attractions that will no doubt lead to a new generation of special-purpose IR spectrometers within a few years. The principal systems used at present are semiconductor diode (SD) lasers, spin-flip Raman (SFR) lasers, and parametric oscillators (PO).

Semiconductor-diode lasers (Fig. 2.10) are made from alloy crystals such as $Pb_{1-x}Sn_xTe$, $PbS_{1-x}Se_x$, and $Pb_xCd_{1-x}S$. They may be chemically tailored to yield IR radiation in bands falling in the mid-IR region by fixing the relative concentration of the constituents. Each band is about 40 cm^{-1} wide, but not all frequencies within the band are available. Each continuous segment of the band may be tuned over a 1–2-cm^{-1} range by varying the diode current. The SD chip must be maintained at cryogenic

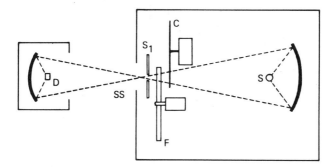

Fig. 2.9 Schematic drawing of dielectric filter spectrometer. Radiation from source S is chopped at C; passes through slowly rotating variable circular filter F, slit S_1, sample space SS; and falls on detector D.

temperatures, and the nominal emission wavelength is selected by varying the diode temperature.

Spin-flip Raman lasers have a somewhat wider tuning range—770–915 cm^{-1} in the case of a CO_2-laser-pumped device [67]. They are tuned by varying the magnetic field in which the device is immersed. They also require cryogenic cooling. Spectral band widths of less than 0.05 cm^{-1} are observed. A balloon-mounted SFR spectrometer has been used to study the concentration of NO in the stratosphere at concentrations in the parts-per-billion (ppb) range [66].

Parametric oscillator devices use laser excitation and are tuned by

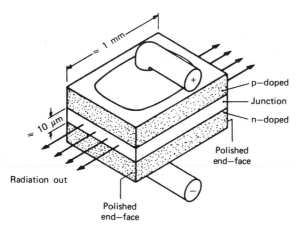

Fig. 2.10 Construction of a p-n junction diode laser [40]. Reproduced with permission of Chapman & Hall, Ltd.

varying their temperature. Their tuning range as well as their spectral line widths are usually broader than those of the SFR and SD lasers [3].

A number of reviews discussing tunable lasers in the IR region are found in the literature [1, 16, 41, 56, 65, 83].

Optoacoustic Devices

Optoacoustic spectroscopy is a related technique in that a tunable laser is used as the source for a substance-specific analyzer. The laser beam, modulated at an acoustical frequency, is directed into a sample chamber in which a sensitive condenser microphone is incorporated into one wall. When the laser frequency corresponds to an absorption line frequency of a gas in the cell, the gas is heated and expands, producing pressure fluctuations at the modulation frequency. These pressure fluctuations are detected by the condenser microphone. The method is extremely sensitive; it can detect a few parts per billion or even lower concentrations under the right circumstances [9, 22, 23, 54, 55].

SPATIAL DETECTION SPECTROMETERS

Spatial detector devices can best be visualized as IR spectrographs having multiple detectors. (Spectrographs, in which all wavelengths are in focus at the detector, are distinguished from monochromators, which are designed so that only the wavelength elements passing the exit slit are free of aberration). One such instrument has been built in which the receiver is a linear array of pyroelectric detectors, which are scanned with an electron beam in a manner analogous to a television camera. The complexity of the receiver makes the instrument expensive and difficult to construct, but it is ideally suited to rapid-scan spectroscopy since it has no moving elements apart from the electron beam.

MULTIPLEX SPECTROMETERS

The term *multiplex* is borrowed from communications theory, where it describes a system of transmitting many sets of information simultaneously over the same channel. Multiplex spectrometers are attractive because they are able to utilize the IR energy much more efficiently than the more common sequential dispersive spectrometers. The multiplex advantage or Fellgett's advantage [24] arises because all wavelengths are measured simultaneously; that is, no exit slit (which blocks roughly 99.9% of the radiation) is used. The multiplex advantage can be quantitatively evaluated from a slightly different perspective. In an experimental measurement characterized by random noise, the signal:noise ratio may be increased by repeating the measurement N times. The signal will then increase in proportion to the number of measurements N, but noise will partially average itself out and increase only as $\sqrt{N}$. Thus the net gain is $\sqrt{N}$. In multiplex spectroscopy we are, in effect, making multiple measurements of the intensity of each resolution element and may expect a substantial gain in signal:noise ratio. This logic holds for detectors in which noise is independent of signal level, which it is for thermal detectors but not for photocells, for example [59]. The resulting energy gain may be traded for faster recording speed, better signal:noise ratio, or ability to analyze smaller samples. Decoding the signal is usually a complex operation, however, performed with the aid of a computer.

Dispersive: Hadamard Transform Spectrometers

An optical diagram of a Hadamard transform spectrometer (HTS) is shown in Fig. 2.11. It will be noted that the monochromator is quite

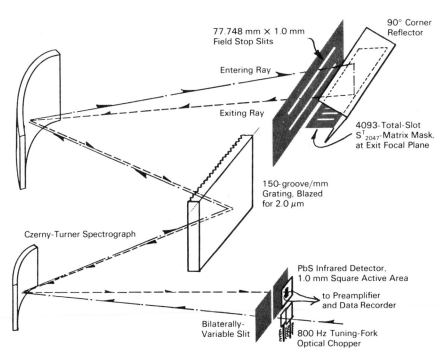

Fig. 2.11 Optical schematic of a 2047-slot Hadamard transform spectrometer using Czerny–Turner optics [21]. Reproduced with permission of The American Chemical Society.

conventional up to the field stop slit. At this point the spectrum is laid out (like an invisible rainbow) along the plane of the aperture mask. Instead of the radiation being limited by an exit slit, however, the complete wavelength distribution passes through the aperture and through the multislit mask, which is fabricated in a precise pattern dictated by the Hadamard transform matrix. Each transparent or opaque mask slot corresponds to a resolution element of the spectrometer. A simplified illustration of a Hadamard mask and the corresponding matrix is shown in Fig. 2.12. In operation, the spectrum is scanned by driving the mask along its horizontal plane by a stepping motor so that each wavelength element successively passes through or is blocked by a "slit" of the mask (Fig. 2.12). Radiation is collected on the receiver and the signal is amplified as usual. In one design, the radiation is "dedispersed" by returning it through the monochromator, and it falls on a detector located in back of the entrance slit (Fig. 2.11). If no sample is present in the sampling space, there is no absorption of radiation and no variation in the signal as the mask scans the exit aperture. If an absorbing sample is present, the detector sees a change in signal when the wavelength element corresponding to an absorption band passes an open slot in the mask,

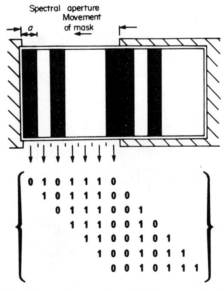

Fig. 2.12 Simple Hadamard encoding mask containing seven elements and its corresponding Hadamard matrix: a is element width. The matrix is presented in oblique form to demonstrate its relationship to the mask. Reproduced, with permission, from *Spectrochimica Acta* [53].

but none when the wavelength element impinges on a blocking slot. The detector output, which corresponds to the sum of all the wavelengths passed by the spectrometer, is stored and later processed to decode the data in terms of an intensity/wavelength plot (i.e., a spectrum) [64]. The system is called a "Hadamard transform" spectrometer because the exit focal plane mask is designed to conform to the Hadamard matrix [77]. If the spectrometer has N resolution elements (e.g., 2000 units at 1-cm^{-1} resolution, 2000–4000 cm^{-1}), the record of the scan has the effect of creating a set of N simultaneous equations that are subsequently solved to define the spectrum.

Advantages of the Hadamard transform spectrometer include the multiplex advantage; that is: a much higher energy throughput than a grating spectrometer (numerically equal to $\sqrt{N}/2$) since all wavelengths are being observed at all times, the ability to record only the wavelengths of interest, and the use of well-proven technology of dispersive spectrometers. Disadvantages include the need for a computer to decode the data and, in practice, errors in coding due to optical aberrations and lack of temperature control of the mask. Also, grating efficiency is high only over a relatively limited region so coverage of the entire spectrum is impractical.

A 255-slot HTS showed the predicted eightfold gain in signal:noise ratio [20]. A 2047-slot HTS has also been built [21], but no performance data were reported.

It has been claimed that the HTS is inherently superior to the interferometer spectrometer, but Hirschfeld and Wyntjes [44] have pointed out that, in fact, the reverse is true. This conclusion has been verified by Meaburn [61], who has calculated relative performance for eight types of spectrometers.

The HTS does have a domain of strength in the IR, however; for spectrometry of selected short spectral segments it gives the multiplex advantage but avoids some of the drawbacks of the IR interferometer.

Nondispersive: Interferometer Spectrometers

The interferometer spectrometer has some important advantages over dispersive spectrometers, some of which are inherent in the design and others of which accrue because the output data are stored in digital form in a computer memory. The basic design is rather simple (Fig. 2.13). The interferometer consists of one fixed and one movable mirror and a beamsplitter. An IR source and a detector, along with appropriate transfer optics, complete the spectrometer.

The operation of the instrument can easily be understood by reference

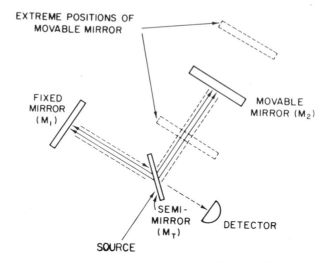

Fig. 2.13　Optical schematic of an IR interferometer. Courtesy Block Engineering, Inc.

to Figs. 2.13 and 14. (More detailed explanations are given elsewhere [36, 37, 45, 58]). Consider the situation if the source is monochromatic. Half of the source radiation is reflected by the beamsplitter, goes to M_1, and back through the beamsplitter to the detector. The other half of the radiation is transmitted by the beamsplitter, goes to M_2 and back to the

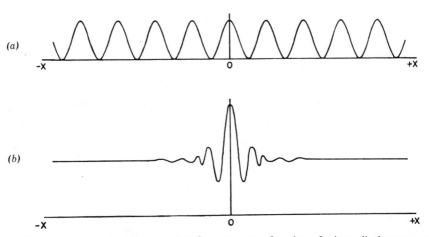

Fig. 2.14　Detector output from an interferometer as a function of mirror displacement x for: (a) monochromatic source and (b) idealized polychromatic source. At position 0 both arms of the interferometer are of equal length. Reproduced, with permission, from *Applied Spectroscopy* [45].

beamsplitter, where a portion of it is reflected and also reaches the detector. When the two arms of the interferometer are of equal length, the two beams interfere constructively. If M_2 is displaced in either direction by the distance $\lambda/4$, the path of the radiation in that arm is changed by $2\lambda/4$ or $\lambda/2$, and the two beams interfere destructively. As M_2 moves another increment of $\lambda/4$, constructive interference again occurs, and so on. The difference in the optical path between the two arms of the interferometer is called the *retardation*. The maximum retardation of an IR interferometer may typically be 2–21 cm. As M_2 moves, the detector sees a beam in which the intensity varies as the cosine wave shown in Fig. 2.14a. If $I(x)$ is the intensity of the beam at the detector, x is the mirror displacement in centimeters and $B(\nu)$ represents the intensity of the source as a function of frequency ν in cm^{-1}, the equation for the signal is

$$I(x) = B(\nu) \cos(2\pi x \nu) \tag{2.2}$$

If we now add a second monochromatic source having a different frequency, the resulting plot of mirror position/intensity will look exactly like that of Fig. 2.14a except that the peaks will crest at different positions on the abcissa. The detector will see the sum of the two cosine waves, or

$$I(x) = B(\nu_1) \cos(2\pi x \nu_1) + B(\nu_2) \cos(2\pi x \nu_2) \tag{2.3}$$

If we add a third, a fourth, and so on, up to an infinite number of frequencies (i.e., polychromatic source), the detector will respond to the sum of all the cosine waves:

$$I(x) = \int_{-\infty}^{\infty} B(\nu) \cos(2\pi x \nu) d\nu \tag{2.4}$$

This signal will be essentially constant over most positions of the movable mirror, but when the two arms of the interferometer are of equal length, all the cosine waves are in phase and the interferogram will show a "center burst" (Fig. 2.14b). (In practice, no source is perfectly polychromatic, and some atmospheric absorptions are always present. The interferogram will thus show some "wiggles" in the wings).

We can use similar logic to understand the appearance of the interferogram (and its decoding) when absorptions are present in the spectrum. Suppose that we now *subtract* one frequency from the polychromatic source (i.e., interpose an absorption band). An inverted cosine wave of that frequency will appear across the interferogram. Subtracting a second frequency will result in subtraction of a second cosine wave, and so on. The resulting interferogram will then be a synthesis of all frequencies except those specifically absorbed by the sample. A typical interferogram and resulting single-beam spectrum for polystyrene are shown in Fig. 2.15.

Equation 2.4 is one-half of a cosine Fourier transform pair. The other equation is

$$B(\nu) = \int_{-\infty}^{\infty} I(x) \cos{(2\pi x\nu)}dx \tag{2.5}$$

These two equations define the relationship between the interferogram and the spectrum. The transformation of the interferogram into a spectrum is done by a computer, using the Cooley–Tukey transform algorithm [18, 25]. At 0.5 cm^{-1} resolution, approximately 16,000 computer words are used to calculate a 400–3800-cm^{-1} spectrum.

Because the interferogram cannot in practice extend from $+\infty$ to $-\infty$, the interferogram is truncated using a suitable mathematical function. Boxcar truncation such as shown in Fig. 2.16b gives, on Fourier transformation of a cosine wave, a resolution function such as that shown in Fig. 2.16d. The side lobes (or feet) can be suppressed by multiplying the interferogram by an *apodization* (literally, without feet) function such as a triangular function (Fig. 2.17b). Other apodization functions are used also; resolution, band shape, and photometric accuracy of the absorptions depend in part on the apodization function chosen [2, 14, 35, 36].

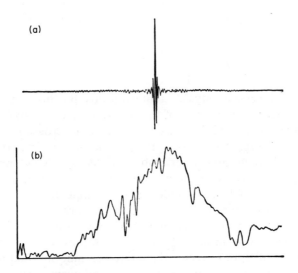

Fig. 2.15 (*a*) Interferogram of blackbody source with polystyrene absorption; (*b*) resulting single-beam IR spectrum (46). Reproduced with permission of American Chemical Society.

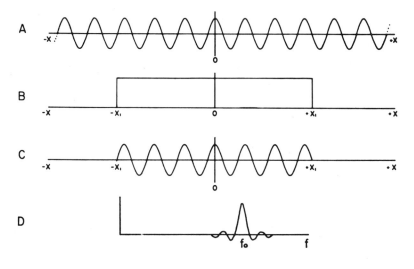

BOX CAR APODIZATION

Fig. 2.16 Effect of boxcar truncation: A, infinitely long cosine wave; B, truncation function; C, cosine wave resulting from multiplication of A and B; and D, pictorial Fourier transform of truncated cosine wave. Reproduced, with permission, from *Applied Spectroscopy* [45].

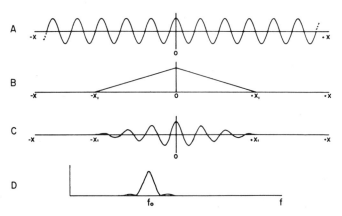

Fig. 2.17 Effect of triangular apodization: A, cosine wave; B, apodization function; C, apodized cosine wave resulting from the multiplication of A and B; and D, pictorial Fourier transform of the apodized cosine wave. Reproduced, with permission, from *Applied Spectroscopy* [45].

31

A simplistic approach to calculating the energy gain of the interferometer over the dispersive spectrometer gives a truly astounding result. We have already discussed Felgett's advantage, $\sqrt{N}$, where N is the number of resolution elements ($N = 3400$ and $\sqrt{N} = 58$ for a 400–3800 cm^{-1} spectrum at 1 cm^{-1} resolution). Additionally, the interferometer has greater throughput—a gain called *Jacquinot's advantage* (after the French scientist Jacquinot) and can be thought of as arising because the interferometer has a circular entrance aperture rather than an entrance slit. Thus advantages in energy of 80–200 times have been claimed for the interferometer.

In practice, however, the observed advantage is much smaller—on the order of 3–10 in the signal:noise ratio over a grating spectrometer. The reasons for this discrepancy have been systematically explored by Griffiths, Sloane, et al. [38]. They find that the advantage depends in a complex fashion on resolution and wavelength, on detector type, and on other factors. For currently available spectrometers, the major contributor to the discrepancy is the NEP (noise-equivalent power) of the TGS detector, which is 20 times poorer than that of the thermocouple. Also, whereas a well-designed beamsplitter has higher efficiency than a grating, some of the older beamsplitters may be less efficient. Below about 600 cm^{-1}, the efficiency of the Ge/KBr beamsplitter is inferior to that of a grating. The interferometer duty cycle efficiency is less than that of a grating spectrometer. The net result of all these factors is shown graphically in Fig. 2.18. Other similar comparisons have given comparable results [8, 74]. Undoubtedly some of the limitations of interferometers will yield to further developments, and the actual advantage will eventually approach the theoretical one. Further improvements in interferometer performance have been claimed by using a process known as "chirping," that is, putting a dispersive material in one arm of the interferometer [72, 73]. The central peak of the interferogram is thus spread. Computations are more complex, but an increase in signal:noise ratio can be achieved.

Because of its unique characteristics, Fourier transform spectrometry is adaptable to a number of problems that would be difficult or nearly impossible to solve on a grating spectrometer. Some of these applications are described more fully later, but they include problems requiring rapid scanning such as recording "on-the-fly" spectra of GC effluents or gas-phase kinetic data.

An apparent advantage of Fourier transform spectrometry arises from the fact that the spectral data are stored in digital form in a computer memory and thus may be easily manipulated; noise can be reduced by repetitive scanning and signal averaging, spectra can be scaled to a factor,

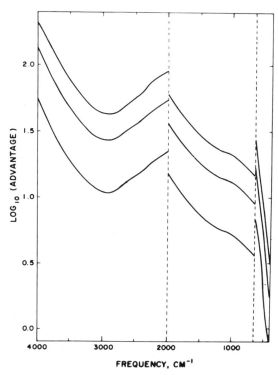

Fig. 2.18 Calculated total advantage (on a logarithmic scale) of Fourier transform spectrometer (similar to Digilab FTS-14) over grating spectrometer (similar to Beckman 4240). Double-beam transmission spectra are assumed at resolutions of 2 cm^{-1} (upper curves), 4 cm^{-1} (middle curves), and 8 cm^{-1} (lower curves). Reproduced, with permission, from *Applied Spectroscopy* [38].

ratioed to or subtracted from another spectrum, and so on. However, grating spectrometers equipped with digital data-processing devices can undertake these same operations [10, 60]. Some of these "tricks" require excellent wavelength repeatability—on the order of 10^{-3} cm^{-1} [42, 43].

Among the drawbacks of Fourier transform spectrometry are the need for highly precise, and thus expensive, interferometer components; for instance, the tilt of the moving mirror must not change by more than one-half wavelength during a scan [34]. Also, the necessity of a computer to transform the interferogram and the difficulty of troubleshooting malfunctions may be obstacles to spectroscopists accustomed to dispersive spectrometers. The spectral range, while adequate, is limited (400–3800 cm^{-1} is typical) and performance deteriorates (i.e., the spectrum becomes noisy) near the limits of the range because of reduced beam-splitter efficiency. A different spectral range requires a different beamsplitter. The entire spectrum is always scanned, and each wavelength has equal time weight; in a grating spectrometer, the use of speed suppression allows more rapid scanning in sparsely absorbing regions, or regions of no interest can be omitted from the scan. A spurious electrical signal or a missed sampling point can have a marked effect on the spectrum by distorting band shapes and causing loss of resolution. A phenomenon known as "aliasing" or "folding" can produce spurious spectral absorptions if the proper optical or electronic filtering is not used [46]. A good discussion of Fourier transform IR spectroscopy is given in the monograph by Griffiths [36].

SELECTION OF A SPECTROMETER

A great variety of commercial instrumentation is available, and choice of a particular unit is influenced primarily by the potential uses for the spectrometer. Other considerations include the availability and quality of service and the personal preferences of the individual who will supervise operation of the spectrometer.

If the instrument is intended for general qualitative and quantitative analyses, as well as for molecular structure research, for example, a double-beam spectrometer of a fairly high degree of sophistication might be selected. If, on the other hand, it is to be used by organic chemists for characterization much as a melting-point apparatus might be, minimum flexibility is desirable. For routine quantitative analysis, low-resolution–high-ordinate-accuracy "IR colorimeters" with built-in computational capabilities are available at minimum cost. If a variety of sophisticated work is to be done, an interferometer spectrometer should be considered. Griffiths [34] lists three situations where IR interferometry is significantly superior to dispersive spectroscopy: (1) applications where a large number of resolution elements is needed (e.g., high-resolution spectroscopy over a wide spectral range), (2) rapid scan applications, and (3) applications where the signal is so weak that an unacceptably long time is needed to measure the spectrum conventionally (e.g., highly absorbing samples or very weak absorptions). It can also improve productivity markedly when large numbers of samples requiring minimum preparation are to be analyzed. The interferometer has no particular advantage for simply recording spectra, however, if a variety of samples requiring different preparation techniques are encountered. Nor does it have any advantage if only limited spectral regions are to be scanned routinely. In fact, under such circumstances the dispersive spectrometer is preferred [8].

In a new IR installation, uses of the spectrometer usually expand rapidly beyond the areas envisioned initially. Therefore, assuming cost is not an overriding consideration, it is usually wise to purchase an instrument with flexibility somewhat in excess of current requirements.

PERFORMANCE CRITERIA

Possession of a good IR spectrometer is no assurance that good spectra will be obtained from it. Even if it is assumed that the instrument is adjusted to give peak performance, the technique of the operator is of overwhelming importance in determining the results. It is safe to say that perhaps 90% of all IR spectrometers (except for the very simplest bench-top models) are not used at their full potential. Before discussing proper use of the instrument controls, however, we define some specialized terms that are important for understanding the following sections.

Definitions

Noise. The random variations in the signal constitute noise. It is measured as the root-mean-square (RMS) deviation of the recorder pen from its average position over a period of time. The RMS noise level is approximately 25% of the peak–peak noise (see Fig. 2.19). Noise arises

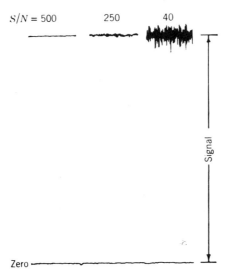

Fig. 2.19 Signal:noise ratios for several noise levels.

from (1) Johnson noise, or thermal motion of electrons in the receiver element, (2) statistical thermal variations within the element, (3) random motion of electrons within the cables and other components of the amplifying system, (4) spurious electrical signals from defective amplifier components or from poor amplifier design, and (5) electrical signals originating outside the instrument. For a well-designed instrument in good adjustment, the noise at the recorder will be largely Johnson noise, which is given by the equation

$$(e_J)^2_{av} = 4kTR(\Delta f) \times 10^{-7} \qquad (2.6)$$

where k is the Boltzmann constant; T is the temperature, R is the detector resistance; and (Δf) is the width of the frequency band passed (which varies inversely with the response time).

Resolution. This is the smallest frequency or wavelength interval distinguishable under given conditions. In a dispersive spectrometer it depends on the resolving power of the dispersing element, the slit width, and the design and adjustment of the optical system. In an interferometer spectrometer, it depends on the maximum retardation, the amount of beam divergence, the apodization function, and, of course, the quality of the optics.

Time constant. For a critically damped system, this is the time required for an element to reach $1/e$ of its ultimate response on instantaneous exposure to constant radiation (where e is the base of natural logarithms).

Response time. Recorder pen response time usually means the time required for full-scale pen travel under slightly underdamped conditions. It is numerically equivalent to about four time constants.

Signal:noise ratio. This is the ratio of the full-scale deflection to the RMS noise level (Fig. 2.19).

Slit width (mechanical). This is the actual opening of the slits, measured in micrometers or millimeters.

Slit width (spectral). This is the range of frequencies transmitted by the monochromator through the exit slit, measured at one-half the maximum intensity. The spectral slit width can be calculated from the mechanical slit width of a dispersive spectrometer [13], provided that certain parameters characteristic of the spectrometer are known.

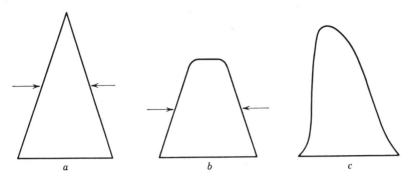

Fig. 2.20 Some possible spectrometer slit functions. The mechanical slit width is indicated at half height of energy peak.

Slit function. This is the intensity/frequency curve for radiation emerging from the exit slit where the entrance slit is illuminated with monochromatic radiation. Some idealized slit functions are shown in Fig. 2.20. The true slit function of a spectrometer can be determined by deconvoluting a narrow absorption line of known Doppler shape [48]. This method avoids errors encountered in determining the slit function by scanning the emission of a coherent source such as a laser. Improperly aligned spectrometers may give asymmetric slit functions (Fig. 2.20c) or even patterns showing structure. The observed band shape is the convolution of the slit function and the natural band shape (Fig. 2.21).

Band width. This is the width of an absorption band, in terms of wave number or wavelength, measured at one-half its peak absorbance value

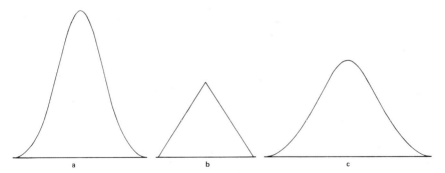

Fig. 2.21 Effect of the finite monochromator slit on band shape: (*a*) true band shape; (*b*) slit function; (*c*) observed band shape.

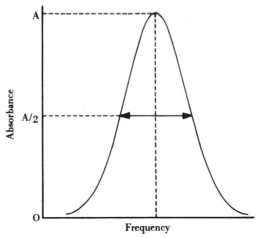

Fig. 2.22   Absorption band width, measured at half peak absorbance.

(Fig. 2.22). It is sometimes called the *half-band width,* or, more properly, the *half-height band width.*

Zero. This is the pen position reached when the energy in the sample beam is blocked. Each wavelength has a true zero point and a false zero, which differs from the true zero by the amount of stray radiation.

Stray or scattered radiation. This is radiation of unwanted frequencies reaching the detector. Among the more common causes of stray radiation are scratches, smudges, or minute imperfections in the mirrors or dispersive element, unwanted orders from gratings, surface reflections within the monochromator, warm slits or warm samples, and warm or dirty choppers.

Approximate measurement of stray radiation may be carried out in several ways. One method uses a heavy layer of the sample so that the band being measured shows virtually total absorption. This method is not exact since the sample itself absorbs some short-wavelength radiation that would otherwise contribute to the stray energy. Or, some material that has a very strong sharp absorption band at the frequency of interest (Table 2.1*a*) may be scanned. One common method utilizes a window (glass, quartz, LiF, CaF_2, NaCl, or KBr) that transmits higher-frequency radiation (the source of most of the stray light) but cuts off at frequencies

TABLE 2.1a Absorption Bands Suitable for Determination of Stray Radiation[a]

Wavenumber Range	Wavelength Range	Substance[b]	Path (in mm)
3600–3620	2.76–2.78	Phenol	20.0 (0.05 M solu-
3260–3335	3.00–3.07	Phenylacetylene	tion) 0.5
3045–3075	3.25–3.28	Methylene chloride	1.0
2380–2460	4.07–4.20	Chloroform	10.
1675–1760	5.68–5.97	Acetone	0.5
1420–1590	6.30–7.05	Carbon disulfide	1.0
1190–1240	8.07–8.42	Chloroform	1.0
1120–1125	8.90–8.93	Tetrachloroethylene	1.0
880–940	10.6–11.4	Tetrachloroethylene	1.0
740–800	12.5–13.5	Carbon tetrachloride	0.1
645–700	14.3–15.5	Benzene	0.5

[a] Data from Kartha [52].
[b] All measurements made on undiluted liquids except phenol.

slightly higher than the band being studied (Table 2.1b). The difference in pen deflection obtained between a metal shutter and the partially transmitting window gives a measure of the stray light.

Stray radiation is minimized in commercial spectrometers by careful design and placement of components and by judicious use of long-wavelength pass dielectric filters. Below 600 cm^{-1} the problem becomes progressively more severe, as the ratio of unwanted to wanted energy becomes very large.

TABLE 2.1b Window Materials for Determination of Stray Radiation[a]

Wavenumber Range	Wavelength Range	Substance	Thickness (in mm)
1000–2000	5.0–10.0	SiO_2 (quartz)	7
700–1200	8.5–16.0	LiF	7
500–900	11.0–20.0	CaF_2	7
250–500	20.0–40.0	NaCl	14
250–300	33.0–40.0	KBr	14

[a] Data from Perkin-Elmer Corp., Model 521 Instruction Manual, July 1963.

Optimizing the Spectrometer Variables

It may be helpful to recall that the performance of an IR spectrometer is limited by available energy, and all adjustments of operating parameters are made to attain the optimum compromise of scanning variables. These conditions depend on the use to be made of the spectrum and the nature of the sample. Conditions that are best in one situation may not be suitable under other circumstances.

When any appreciable number of spectra are to be accumulated and compared, a set of standard schedules should be established for the spectrometer so that all spectra can be run under reproducible conditions. It may seem like a waste of time to run routine samples carefully and reproducibly, but experience has shown that many "routine" spectra unexpectedly end up in the reference library. Even if they do not, more useful information can be obtained from a good spectrum than from a carelessly run one. Spectrometer "survey" conditions should be used only for checking sample thickness; it is best never to record at survey speed. The following discussion of "trading rules" applies to dispersive spectrometers. Interferometer spectrometers also have trading rules, which are discussed later.

It is important for users of dispersive spectrometers to understand the nature and use of the three basic parameters affecting the operation of the spectrometer. Even where automatic programs for adjustment of the variables are provided, the user should still understand which parameters are being changed and how the change will affect the spectrum.

The three basic variables available to the spectroscopist are resolution (slit width), noise level (gain), and scan speed (response time). Any two of these may be arbitrarily chosen; the third parameter is then fixed by that choice. (Throughout this discussion it is assumed that we wish to minimize distortion in the spectrum.) Although the discussion is directed toward operation of double-beam optical null spectrometers, ratio-recording instruments (which do not use an optical attenuator) are subject to the same limitations of the basic variables.

As we have seen earlier (pp. 15–19), optical null spectrometers (the most common type as of this writing) use an electromechanical servo system to drive the recorder pen. The servo system consists of a closed loop containing the comb or optical attenuator in the reference beam, thermocouple or other detector, amplifier, and a servo motor that drives the attenuator in or out of the reference beam. The gain setting of this servo loop is critical to proper operation of the spectrometer. If the gain is too low, the system will respond sluggishly and incompletely; if it is too high, the attenuator servo motor (and the pen) will overshoot badly or even break into oscillation. Damping of the servo system may be changed by adjusting the response time; thus the response time and gain must be matched to give the proper servo response. The scan speed adjustment must be consistent with the response time. Nothing is gained and much is lost by scanning more rapidly than the pen can respond.

The servo loop energy is affected by the slit width as well as by the servo amplifier gain setting. These two adjustments must be balanced to keep the energy of the servo-loop constant. A higher gain setting is used with a narrow slit, and vice versa. Of course, using a higher gain setting also results in a higher noise level in the spectrum, and a narrow slit results in better resolution.

The quantitative relationship between the signal:noise ratio, S/N, scan time, t, and slit width, w, is

$$S/N = ct^{1/2}w^2 \tag{2.7}$$

where c is a constant.

This equation comes about as follows. Equation (2.6) for Johnson noise may be rewritten:

$$N \propto (\Delta f)^{1/2} = \text{(time constant of servo loop)}^{-1/2} \tag{2.8}$$

Furthermore, since the signal is proportional to the square of the slit width, or $S \propto w^2$, we may combine these relationships to obtain equation (2.7).

Equation (2.7) also tells us that the signal:noise ratio increases with the square of the slit width. Doubling the slit width will decrease the noise level by a factor of 4 (provided that the gain of the servo system is decreased to compensate for the larger amount of energy reaching the detector).

Noise also can be decreased by increasing scan time. Putting equation (2.7) another way, we have

$$\text{Resolution} \propto 1/w = c't^{1/4}(S/N)^{-1/2} \tag{2.9}$$

Thus to double the resolution by halving the slit width, we must increase the scanning time by a factor of 16. It is clear that if the spectrum is scanned slowly enough so that the servo system can follow the signals, an appreciable gain in resolution can be achieved only at an impractical increase in scan time.

Since doubling the resolution doubles the number of spectral (resolution) elements, it has also been argued that scan time should be increased by 32 rather than by 16 [74]. However, depending on one's definition of "resolution element," the way in which the measurement is made, and the width of the absorptions being measured, either factor can be correct.

Digitized IR spectra can also be smoothed by computer manipulation of data, using a moving-point polynomial averaging function [85]. If the data are hidden in the noise, repetitive scanning techniques with successive accumulation of data can be used to increase the signal:noise ratio (see p. 107).

It is appropriate at this point to discuss scanning speed in relation to the time constant of the electronics and the response speed of the servo system. The purpose of the band-pass filter in the amplifier circuit (which passes signals of the chopping frequency and rejects all other signals) is to reduce random electrical signals (noise) and 60-cycle pickup from power lines. The more sharply the filter is tuned, the lower the noise level but, also, the slower the response time of the system will be. In addition, some circuits incorporate additional filters for electronic averaging of noise. These parameters ordinarily constitute the limiting factors in determining the scan speed since the response speed of the pen-drive motor and associated circuitry is usually less than the amplifier response time.

It is possible to make quite large scanning errors in both intensity and position, particularly of sharp narrow bands. According to Stewart [80], "A scanning speed of less than 0.4 observed band-widths per response period is required in order to observe intensities in error by less than 2%. This is an experiment which spectroscopists should perform for themselves now and then as a reminder that it is very easy to scan too fast." Widths observed for a number of randomly chosen absorption bands are shown in Fig. 2.23.

With this background, we are now in a position to establish "general-purpose" conditions for the operation of the spectrometer. We outline the procedure briefly and then show how the basic parameters are systematically changed to meet special purpose requirements such as high resolution, rapid scanning, or high quantitative accuracy. A more complete discussion of the process is given by Potts and Smith [69].

It is assumed that the spectrometer components are functioning normally and that the spectrometer is operating within the manufacturer's specifications.

First we (arbitrarily) choose a slit-width program. This may be the setting suggested by the manufacturer, or (preferably) one that is 1.2–1.5 times wider. If speed suppression is available, we may choose 1 h (60 cm^{-1}/min) as the scan time (this time will be shortened later). If not, 0.5–0.6 h for the spectrum is reasonable. The servo response must be consistent with the scan time; to minimize distortion, no more than 0.4 band widths per response period should be scanned. Since most band widths in liquids and solids are at least 3 cm^{-1} (Fig. 2.23), a pen response of

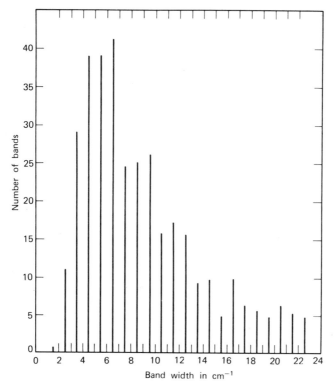

Fig. 2.23 Distribution of band widths of 347 liquid and solid sample absorption bands, scanned on spectrometer with spectral slit width of 1 cm⁻¹.

about 1-sec full scale is proper for a 1-h scan time. The gain setting is now adjusted to give the correct energy in the pen servo system. At the proper setting, the pen will overshoot about 2–4% in response to a sudden partial blocking or unblocking of the sample beam (this test should be carried out in a region of no atmospheric absorption, e.g., at 1000 cm⁻¹). The noise level is now checked. A *small amount of noise* (say, ±0.2%) *is desirable*, as an indicator that the spectrometer is operating at its maximum efficiency. A smooth trace may indicate a dead servo system. Excess noise is also undesirable, as it interfers with the operation of the automatic speed suppression circuitry and distorts the spectrum.

If the noise level is too high or too low, the slit is readjusted and the cycle repeated until the results are satisfactory.

Pen drift or balance is now set so that the pen shows no consistent motion when both beams are completely blocked. Automatic speed suppression is then added by scanning with both beams blocked, increasing the suppression control until the scan speed slows (the instrument suppresses on noise), and then reducing the suppression slightly. The scan time can now be shortened by a factor of 2 or 3; while scanning an absorption band, the spectrometer should automatically reduce its scan speed to that originally chosen (1 cm^{-1}/s in the preceding example). Further slight adjustment of the suppression control may be necessary to achieve the proper "suppressed" speed.

The next step, which is absolutely essential, is to verify that the conditions chosen are indeed correct. A sample that has a good variety of sharp and broad bands, such as indene [12, 51] is scanned at normal speed and rescanned at 25% normal speed. The two scans should be essentially identical. If they are not, the controls will have to be fine-tuned to arrive at acceptable performance.

Once standard conditions are determined, performance of the spectrometer should be verified frequently (once a day in a busy laboratory) by running a spectrum of a test material (see pp. 48–50). Some minor adjustment of gain and scan time may be necessary from time to time; the slit program, once established, should be strictly maintained constant.

Using equation (2.7), we can derive proper values for the variable parameters to meet special conditions. Some equivalent relationships between the variables are shown in Table 2.2. By way of example, suppose we wish to scan a portion of the spectrum with the ordinate scale expanded by a factor of 10. Unless we make some compensating adjustments, the noise will also be increased by 10 times and no advantage will be gained. We can open the slit by a factor $\sqrt{10}$, decrease the gain to give the original noise level, and scan at our usual speed. Or, we can increase the response time and the scan time by a factor of $(10)^2 = 100$, or use a combination of factors. We might, for example, open the slit

TABLE 2.2   Variations of Spectrometer Parameters for
Equivalent Servo-loop Energies

Condition	Noise	Slit	Scan Time (Response)
Normal	1	1	1
Low noise	0.25	2	1
Low noise	0.25	1	16
High resolution	1	0.5	16
High resolution	2	0.5	4
Rapid scan	1	$\sqrt{2}$	0.25

by a factor of 2, increase the scan time by a factor of 4, and accept a slightly higher than normal noise level. Use of the trading rules to adapt to other special needs is summarized in Table 2.3.

Trading rules for interferometer spectrometers have been derived and experimentally verified by Griffiths [33]. For *any* type of spectrometer, the signal:noise ratio of a spectrum measured at a given resolution is proportional to the square root of the measurement time. In rapid-scan interferometers, scan time is increased by using additional scanning cycles and signal averaging either the resultant interferograms or the spectra (usually the former). In slow-scan interferometers, such as are commonly used for far IR work, the velocity of the moving mirror can be reduced, but this change requires adjustment of the noise-filter time constant [52]. The resolution of an interferometer is changed by changing the retardation (length of travel of the moving mirror). The interferometer optics are generally throughput matched to some retardation to give optimum performance at a reasonable resolution (e.g., 2 cm^{-1}). For measurements at lower resolutions, the maximum throughput allowed theoretically cannot be reached in practice because of limitations imposed by the size of the source or the detector. Under these conditions, the throughput is not changed as the resolution is varied (the constant throughput case). At

TABLE 2.3 Summary of Procedures for Establishing Operating Parameters for Special Purpose Spectra[a]

Low noise
    Establish general-purpose conditions
    Determine noise-level reduction factor
    Widen slits by (factor)$^{1/2}$ and reduce gain by (factor) or increase response and scan time by (factor)^2
    Run test spectrum
Limited energy
    Establish general-purpose conditions
    Determine energy attenuation factor
    Widen slits by (factor)$^{1/2}$ or increase gain by (factor) and increase response time and scan time by (factor)^2
    Run test spectrum
High resolution
    Establish general purpose conditions
    Determine slit-reduction factor
    Restore servo energy by increasing gain
    Increase response time and scan time to obtain tolerable noise level
    Run test spectrum

---

[a] Reprinted, with permission, from *Applied Optics* [69].

some point, however, as resolution is increased, beam divergence must be reduced in order to benefit from increased retardation (the variable throughput case); this operation is usually accomplished by changing the diameter of an aperture placed at a focal point in the source optics. This aperture serves the same purpose as the entrance slit of a monochromator.

In the *constant throughput* case (source aperture area remains constant), increasing the resolution by increasing the retardation by a factor of 2 increases the measurement time by 2 and thereby degrades the signal:noise ratio by $\sqrt{2}$. To restore the signal:noise ratio, the measurement time must be doubled again, so the total time increase is 4 (in contrast, a dispersive spectrometer requires a factor of 16 increase in scan time). Conversely, degrading the resolution by 2 yields a gain in measurement time of only 4 with an interferometer, but with a dispersive spectrometer, the gain is 16. Griffiths has tested this rule using an FTS-14 spectrometer at resolutions of 2 cm^{-1} and 4 cm^{-1} [33].

At higher resolutions where the beam divergence must be reduced, the *variable throughput* case, the situation is different. To increase resolution by a factor of 2, we must reduce the source aperture by $\sqrt{2}$, which reduces the energy throughput by 2. Recovering the "lost" energy requires an increase by a factor of 2 in measurement time, which increases the noise by $\sqrt{2}$, which in turn requires another factor of 2 increase in measurement time for a total of 4. The retardation is also increased by a factor of 2, which requires a corresponding increase in measurement time of an additional factor of 4. Thus the total increase in measurement time is 16, exactly the same as for a dispersive spectrometer.

The resolution is also affected by the apodization. Spectra having approximately the same signal:noise ratio and resolution may be measured with the same number of scans under two sets of conditions: using a given retardation and triangular apodization or using *half* that retardation and boxcar apodization. Measurement time is halved in the latter case, but sharp lines show side lobes.

In summary, if we start from the point where the interferometer operates at the maximum throughput for low-resolution measurements and trade (degrade) resolution to reduce the scan time, the gain will be less than for the equivalent case in a grating spectrometer, by a factor of 4 in the measurement time. (In other words, the advantage of the interferometer over the grating spectrometer is less at low resolution.) On the other hand, if we need to decrease the throughput to increase the interferometer resolution, the trading rules are exactly the same as those for the grating spectrometer.

Other Factors Affecting Performance

We have seen how the performance of an IR spectrometer is limited by fundamental physical laws. Other, more artificial, limitations may arise from misadjustment of the instrument optics, defective components, or improper environment.

Manufacturers of IR instruments usually send a service engineer to supervise installation and adjust a new instrument to factory specifications. Whereas some slight further gain in performance may still be possible at this point, it is usually not wise to attempt further improvements, since optical adjustments by untrained personnel almost invariably result in performance deterioration. Optical components, if properly set initially, do not change adjustment and rarely need further attention.

Other components, however, and especially vacuum tubes if they are used, occasionally need replacement. Infrared sources gradually deteriorate and require changing periodically. Most manufacturer's manuals give troubleshooting procedures to follow when difficulty is encountered. The procedure used is a systematic isolation of the system (optical, mechanical, or electronic) causing difficulty, then the section (main amplifier, pen, servo, etc.) within the system, and finally the defective component in the section. Erratic operation can also result from intermittent electrical disturbances transmitted through either air or power lines. If the ambient temperature is not relatively constant, the wavelength calibration of the instrument is likely to drift.

A fact often overlooked by the users of double-beam spectrometers is that carbon dioxide and water vapor in the air may become almost totally absorbing in certain regions, particularly at 2320 cm^{-1} and 1400–1700 cm^{-1}, with the result that the pen may become sluggish and respond poorly to optical signals. Rapid scanning in these regions may give an apparently noisy spectrum because the scan speed is appreciable compared to the chopping speed, and the two beams "see" different portions of the atmospheric absorption bands. (A similar effect will be noted if the two beams do not follow exactly the same path through the monochromator or if the amount of scattered light in the two beams is different, possibly due to a smudged source mirror.)

The solution to problems of atmospheric absorption is to purge the spectrometer with dry air or nitrogen. If interference is to be removed completely, some method of enclosing the sample space must be provided and a very rapid flow of purge gas must be used. Fortunately, it is seldom necessary to take such elaborate precautions.

In the far-IR region (beyond 400 cm^{-1}) the problem is more serious because of the strong rotational water vapor bands. In this region careful purging is mandatory and at very low frequencies (10–200 cm^{-1}) evacuation of all or part of the spectrometer is very helpful.

Performance Tests and Spectrometer Calibration

It is strongly recommended that standard instrument settings be established for each type of spectrum run and that these settings be reproduced for subsequent work of the same type. Unbelievable errors in intensities, band shapes, and absorption frequencies can occur if scanning conditions are left to the whim of the spectrometer operator. Misleading spectra obtained under such circumstances can be worse than no spectra at all. The effect of slit width on the intensities of a moderately sharp and moderately broad band is shown in Fig. 2.24.

As spectrometers of higher resolution and greater flexibility are developed, the demands for continuous peak performance of the components become more rigorous. It is important to monitor this performance frequently to prevent unsuspected deterioration in the quality of the spectral measurements, since in even the most reliable spectrometers electronic components can deteriorate, sources warp, and optical surfaces may become smudged or fogged.

Perhaps the most practical performance test is a dynamic one in which the spectrum of a suitable test material is scanned under standard conditions. With the proper choice of standard one can, from a single spec-

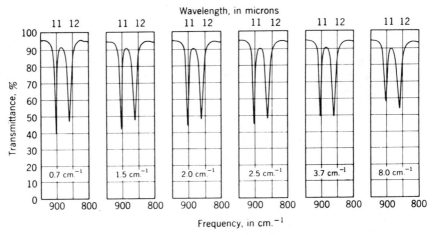

Fig. 2.24   Effect of spectral slit width on absorbance of 861-cm^{-1} and 903-cm^{-1} cyclohexane bands. Reproduced, with permission, from *Applied Spectroscopy* [75].

trum, carry out a daily check on wavelength accuracy, signal:noise ratio, resolution, zero, and reproducibility in absorbance. Determination of stray light and flatness of the I_0 line should also be carried out frequently. Other secondary responses peculiar to the spectrometer used also may require periodic testing.

The standard test material ideally should: (1) be stable, (2) be of constant or reproducible thickness, (3) have a large number of absorption bands that fall in all spectral regions and that show considerable variations in absorbance and width, and (4) be readily available. No one substance meets all these criteria, but a number of reasonably appropriate substitutes have been used. Polystyrene film (0.07 mm) is a well-known calibration material in the 600–3400-cm^{-1} region, although it does not have as many bands as are desirable. Gaseous ammonia (200 mm in a 5-cm cell) is useful but somewhat difficult to maintain without pressure change. Potts [68] has recommended a 0.03-mm layer of 1,2-dibromopropane. The use of indene (0.02 mm) has been proposed for calibration, and wavelengths given for 77 absorption bands covering the range 690–4000 cm^{-1} [50, 51] (see Table 2.4a and Fig. 2.25). Pure indene is stable when properly stored [50]. Additional useful bands are provided by the addition of cyclohexanone and camphor (Table 2.4b). The spectrum of this mixture provides a number of convenient points for testing not only wavelength accuracy, but also the other variables mentioned previously.

The calibration of both prism and grating IR spectrometers in the range 600–4300 cm^{-1} has been covered thoroughly in the compilation of spectra and frequencies by the Commission on Molecular Structure and Spectroscopy of the International Union of Pure and Applied Chemistry [15]. It should be noted that any band, unless it is isolated and perfectly symmetrical, will show a position dependence on spectral slit width. It is thus important that the resolution of the spectrometer being calibrated approximately matches that of the instrument on which the spectrum was measured, or that bands showing little sensitivity to spectral slit width be used [15, 71, 82]. These factors have been recognized in the IUPAC tables mentioned earlier.

Transferability of Absorbance Data

Although the peak absorbance repeatability of consecutive runs on an individual spectrometer may be very good (perhaps on the order of 0.3%), when one attempts to run the same sample on another instrument, even of the same make and model (or on the same instrument at another time), the agreement may be no better than 10–20%. Obviously, then, a good deal of restraint must be exercised in quoting probable analytical errors, unless care is taken that each analysis be standardized on one particular instrument.

The reasons for this state of affairs are several. First, if slit width, gain, response time, and scan speed are not matched properly, results even on the same instrument will be erratic. Second, the effective slit width must be less than 20% of the band width to give quantitative extinction coefficients [5], and the use of narrow slits is not compatible

TABLE 2.4a   Absorption Maxima Recommended for Calibration Purposes[a,b]

Band	$\nu_{vac} - cm^{-1}$	Cell (in mm)	Band	$\nu_{vac} - cm^{-1}$	Cell (mm)
1	3927.2 ± 0.56	0.2	41	1885.1 ± 0.42	0.2
2[c]	3901.6 ± 0.64	0.2	42	1856.9 ± 0.52	0.2
3[d]	3798.9 ± 0.86	0.2	43[d]	1826.8 ± 0.56	0.2
4[d]	3745.2 ± 0.72	0.2	44	1797.7 ± 0.50	0.2
5[d]	3660.6 ± 0.98	0.2	45	1741.9 ± 0.50	0.2
6		0.2	46[d]	1739.2 ± 0.78	0.2
7		0.2	47	1713.4 ± 0.66	0.2
8[d]	3297.8 ± 1.06	0.2	48[d]	1684.9 ± 1.14	0.2
9[c]	3139.5 ± 0.44	0.2	49[d]	1661.8 ± 0.64	0.2
10[c]	3110.2 ± 0.44	0.2	50	1609.8 ± 0.42	0.025
11[d]	3068.9 ± 0.66	0.025	51	1587.5 ± 0.26	0.2
12[c]	3025.4 ± 0.26	0.025	52[d]	1574.5 ± 0.62	0.2
13[c]	3015.3 ± 0.52	0.025	53	1553.2 ± 0.20	0.2
14[c]		0.025/0.2	54[c]		0.2
15[d]	2887.6 ± 0.82	0.025	55[d]	1457.3 ± 0.38	0.025
16[c]		0.2	56[d]	1393.5 ± 0.76	0.025
17	2770.9 ± 0.44	0.2	57	1361.1 ± 0.16	0.025
18[c]		0.2	58[d]	1332.8 ± 0.42	0.025
19	2673.3 ± 0.56	0.2	59	1312.4 ± 0.18	0.025
20[c]	2622.3 ± 0.24	0.2	60	1288.0 ± 0.08	0.025
21	2598.4 ± 0.16	0.2	61[c,d]	1264.0 ± 0.12	0.025
22		0.2	62	1226.2 ± 0.28	0.025
23	2525.5 ± 0.32	0.2	63	1205.1 ± 0.20	0.025
24[c]		0.2	64	1166.1 ± 0.08	0.025
25[c]		0.2	65[c]		0.025
26[d]	2439.1 ± 0.24	0.2	66	1122.4 ± 0.32	0.025
27[c]		0.2	67[c]		0.025
28	2305.1 ± 0.42	0.2	68	1067.7 ± 0.30	0.025
29[c]	2271.4 ± 0.08	0.2	69	1018.5 ± 0.32	0.025
30[c]	2258.7 ± 0.36	0.2	70[c]		0.025
31		0.2	71	947.2 ± 0.36	0.025

TABLE 2.4*a* (Continued)

Band	$\nu_{vac} - cm^{-1}$	Cell (in mm)	Band	$\nu_{vac} - cm^{-1}$	Cell (mm)
32		0.2	72	942.4 ± 0.38	0.025
33	2172.8 ± 0.30	0.2	73	914.7 ± 0.16	0.025
34	2135.8 ± 0.68	0.2	74	861.3 ± 0.14	0.025
35	2113.2 ± 0.28	0.2	75	830.5 ± 0.32	0.025
36	2090.2 ± 0.40	0.2	76	765.3 ± 0.22	0.012
37[d]	2049.1 ± 0.82	0.2	77	730.3 ± 0.22	0.012
38[c,d]	2027.0 ± 0.42	0.2	78	718.1 ± 0.24	0.012
39	1943.1 ± 0.52	0.2	79[d]	692.6 ± 0.56	0.012
40	1915.3 ± 0.30	0.2			

[a] Data from Jones and Nadeau [50, 51].
[b] Variation given represents twice standard deviation; omitted band positions have been found unreliable for calibration.
[c] This band may not be resolved by the smaller types of prism spectrometers.
[d] Because of asymmetry, superposition on atmospheric water vapor or carbon dioxide bands, or for other reasons, these bands are less suited for accurate calibration.

TABLE 2.4*b*   Equimixture of Indene, Camphor, and Cyclohexanone; Wave Number (Vacuum)[a]

Band No.	Wavenumber (in cm^{-1})
1	592.1
2	551.7
3	521.4
4	490.2 ± 1
5	420.5
6	393.1
7	381.6
8	301.4

[a] Values are accurate to ±0.5 cm^{-1} unless otherwise indicated. Cell thickness is 0.05 mm. Bands 3 and 8 are from camphor, and band 4 is from cyclohexanone [50].

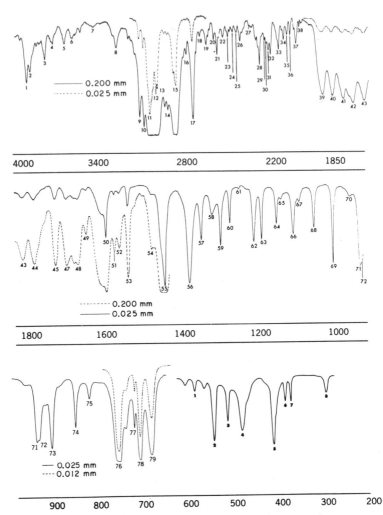

Fig. 2.25 Spectrum of indene (see Table 2.4). Courtesy Perkin-Elmer Corporation.

with the low noise level needed for good quantitative accuracy. The effect of spectral slit width on absorbance is shown in Fig. 2.24. Absorbance values are sensitive to the slit function of a spectrometer. Optical idiosyncracies that result in stray radiation may be important. Third, attenuators used in optical-null double-beam spectrometers are subject to nonlinearity, even when prepared by photoengraving techniques. This nonlinearity is not detected by a Beer's law test [76].

If, then, one wishes to obtain *transferable* peak absorbance data from a spectrometer, he must operate at slit schedules giving spectral slit widths no greater than $0.2w$, where w is the width of the narrowest band to be measured; match scanning speed, slit width, and response time in such a way as to obtain accurate band shapes; and make sure that the optical attenuator (if present) is linear within the accuracy required. This measurement can be carried out by using a rotating-sector attenuator with precisely adjustable blades [49].

It should be noted that the first requirement pertains only if *transferable* absorbance data are required. For most quantitative analyses, the use of *wider* than normal slits to obtain a lower noise level is advantageous (see Chapter 6).

Integrated absorbance data on reasonably isolated absorptions can be reproduced to better than $\pm 2\%$ in different laboratories if spectrometer variables are carefully controlled and a suitable method is used for integration [6].

References

1. Allkins, J. R., *Anal. Chem.*, **47**, 752A (1975).
2. Anderson, R. J., and P. R. Griffiths, *Anal. Chem.*, **47**, 2339 (1975).
3. Baldwin, G. C., *An Introduction to Non-Linear Optics,* Plenum Press, New York, 1969.
4. Bratt, P., W. Engeler, H. Levinstein, A. MacRae, and J. Pehek, *Infrared Phys.*, **1**, 27 (1961).
5. Brodersen, S., *J. Opt. Soc. Am.*, **44**, 22 (1954).
6. Brownlee, R. T. C., D. G. Cameron, B. Ternai, and R. D. Topsom, *Appl. Spectrosc.*, **25**, 564 (1971).
7. Cann, M. W. P., *Appl. Opt.*, **8**, 1645 (1969).
8. Chenery, D. H., and N. Sheppard, *Appl. Spectrosc.*, **32**, 79 (1978).
9. Claspy, D. C., Infrared Optoacoustic Spectroscopy and Detection, in Y.-H. Pao, Ed., *Optoacoustic Spectroscopy and Detection,* Academic Press, New York, 1977.
10. Coates, J. P., *Am. Lab.*, **8**(11), 67 (1976).
11. Coates, V. J., *Spectrochim. Acta*, **15**, 820 (1959).
12. Coblentz Society Board of Managers, *Anal. Chem.*, **38**(9), 27A (1966); **47**, 945A (1975).
13. Coblentz Society Review Committee, *Appl. Spectrosc.*, **11**, 109 (1957).
14. Codding, E. G., and G. Horlick, *Appl. Spectrosc.*, **27**, 85 (1973).
15. Cole, A. R. H., *Tables of Wavenumbers for the Calibration of Infrared Spectrometers,* 2nd ed., Pergamon Press, Oxford, 1977.

16. Colles, M. J., C. R. Pidgeon, *Rep. Progr. Phys.*, **38**, 329 (1975).
17. Conn, G. K. T., and D. G. Avery, *Infrared Methods*, Academic Press, New York, 1960.
18. Cooley, J. W., and J. W. Tukey, *Math. Comput.*, **19**, 297 (1965).
19. Cooper, J., *Rev. Sci. Instrum.*, **33**, 92 (1962).
20. Decker, J. A., Jr., *Appl. Opt.*, **10**, 510 (1971).
21. Decker, J. A., Jr., *Anal. Chem.*, **44**(2), 127A (1972).
22. Dewey, C. F., Jr., *Opt. Eng.*, **13**, 483 (1974).
23. Dewey, C. F., Jr., R. D. Kamm, and C. E. Hackett, *Appl. Phys. Lett.*, **23**, 633 (1973).
24. Fellgett, P., *J. Phys. Radium*, **19**, 187, 237 (1958).
25. Forman, M. L., *J. Opt. Soc. Am.*, **56**, 978 (1966).
26. George, R. S., *Appl. Spectrosc.*, **20**, 101 (1966).
27. Girard, A., *Opt. Acta*, **7**, 81 (1960).
28. Girard, A., *Appl. Opt.*, **2**, 79 (1963).
29. Girard, A., *J. Phys. (Paris)*, **24**, 139 (1963).
30. Golay, M. J. E., *J. Opt. Soc. Am.*, **39**, 437 (1949).
31. Golay, M. J. E., *Rev. Sci. Instrum.*, **20**, 816 (1949).
32. Golay, M. J. E., *J. Opt. Soc. Am.*, **46**, 422 (1956).
33. Griffiths, P. R., *Anal. Chem.*, **44**, 1909 (1972).
34. Griffiths, P. R., *Anal. Chem.*, **46**, 645A (1974).
35. Griffiths, P. R., *Appl. Spectrosc.*, **29**, 11 (1975).
36. Griffiths, P. R., *Chemical Infrared Fourier Transform Spectroscopy*, Wiley, New York, 1975.
37. Griffiths, P. R., C. T. Foskett, and R. Curbelo, *Appl. Spectrosc. Rev.*, **6**, 31 (1972).
38. Griffiths, P. R., H. J. Sloane, and R. W. Hannah, *Appl. Spectrosc.*, **31**, 485 (1977).
39. Hackforth, H. C., *Infrared Radiation*, McGraw-Hill, New York, 1960.
40. Hinkley, E. D., *Opto-electronics*, **4**, 69 (1972).
41. Hinkley, E. D., K. W. Nill, and F. A. Blum, *Top. Appl. Phys.*, **2**, 125 (1976).
42. Hirschfeld, T., *Appl. Spectrosc.*, **30**, 549 (1976).
43. Hirschfeld, T., *Appl. Spectrosc.*, **30**, 550 (1976).
44. Hirschfeld, T., and G. Wyntjes, *Appl. Opt.*, **12**, 2876 (1973).
45. Horlick, G., *Appl. Spectrosc.*, **22**, 617 (1968).
46. Horlick, G., and H. V. Malmstadt, *Anal. Chem.*, **42**, 1361 (1970).
47. Jacquinot, P., *J. Opt. Soc. Am.*, **44**, 761 (1954).
48. Jansson, P. A., *J. Opt. Soc. Am.*, **60**, 184 (1970).
49. Jones, R. N., D. Escolar, J. P. Hawranek, P. Neelakantan, and R. P. Young, *J. Molec. Struct.*, **19**, 21 (1973).

50. Jones, R. N., and A. Nadeau, *Can. J. Spectrosc.*, **20**, 33 (1975).

51. Jones, R. N., and A. Nadeau, *Spectrochim. Acta*, **20**, 1175 (1964).

52. Kartha, B. V., quoted by K. S. Seshadri and R. N. Jones, *Spectrochim. Acta*, **19**, 1013 (1963).

53. Keir, M. J., J. B. Dawson, and D. J. Ellis, *Spectrochim. Acta*, **32B**, 59 (1977).

54. Kreuzer, L. B., *J. Appl. Phys.*, **42**, 2934 (1971).

55. Kreuzer, L. B., N. D. Kenyon, and C. K. N. Patel, *Science*, **177**, 347 (1972).

56. Kuhl, J., and W. Schmidt, *Appl. Phys.*, **3**, 251 (1974).

57. Levinstein, H., *Anal. Chem.*, **41**(14), 81A (1969).

58. Low, M. J. D., and I. Coleman, *Spectrochim. Acta*, **22**, 369 (1966).

59. Marshall, A. G., and M. B. Comisarow, *Anal. Chem.*, **47**, 491A, 1975.

60. Mattson, J. S., *Anal. Chem.*, **49**, 470 (1977).

61. Meaburn, J., *Appl. Opt.*, **14**, 2521 (1975).

62. Moret-Bailly, J., Grille Spectrometers, in K. N. Rao, Ed., *Molecular Spectroscopy: Modern Research*, Academic Press, New York, 1972, p. 327.

63. Moss, T. S., *Infrared Phys.*, **16**, 29 (1976).

64. Nelson, E. D., and M. L. Fredman, *J. Opt. Soc. Am.*, **60**, 1664 (1970).

65. Nill, K. W., *Proc. Soc. Photo-Opt. Instrum. Eng.*, **49**, 56 (1975).

66. Patel, C. K. N., E. G. Burkhardt, and C. A. Lambert, *Science*, **184**, 1173 (1974).

67. Patel, C. K. N., and E. D. Shaw, *Phys. Rev. Lett.*, **24**, 451 (1970).

68. Potts, W. J., Jr., *Chemical Infrared Spectroscopy*, Vol. 1, *Techniques*, Wiley, New York, 1963.

69. Potts, W. J., Jr., and A. L. Smith, *Appl. Opt.*, **6**, 257 (1967).

70. Putley, E. H., *Phys. Technol.*, **4**, 202 (1973).

71. Rao, K. N., C. J. Humphreys, and D. H. Rank, *Wavelength Standards in the Infrared*, Academic Press, New York, 1966.

72. Sheahen, T. P., *Appl. Opt.*, **13**, 2907 (1975).

73. Sheahen, T. P., *Appl. Opt.*, **14**, 1004 (1975).

74. Sheppard, N., R. G. Greenler, and P. R. Griffiths, *Appl. Spectrosc.*, **31**, 448 (1977).

75. Sloane, H. J., *Appl. Spectrosc.*, **16**, 5 (1962).

76. Sloane, H. J., and W. S. Gallaway, *Appl. Spectrosc.*, **31**, 25 (1977).

77. Sloane, N. J. A., T. Fine, P. G. Phillips, and M. Harwit, *Appl. Opt.*, **8**, 2103 (1969).

78. Smollett, M., *Infrared Phys.*, **8**, 3 (1968).

79. Stewart, J. E., *Infrared Spectroscopy: Experimental Methods and Techniques*, Marcel Dekker, New York, 1970.

80. Stewart, J. E., Paper No. 205, Pittsburgh Conference on Analytical Chemistry and Applied Spectroscopy, March 1961.

81. Telfair, W. B., A. C. Gilby, R. J. Syrjala, and P. A. Wilks, Jr., *Am. Lab.*, **8**, (11) 91 (1976).

82. Vorob'ev, V. G., and V. A. Nikitin, *Opt.-Mekh. Prom.*, **38**, 54 (1971).

83. Whiffen, D. H., Lasers in Infrared Spectroscopy, in A. R. West, Ed., *Molecular Spectroscopy*, Heyden, London, 1977.

84. Williams, V. Z., *Rev. Sci. Instrum.*, **19**, 135 (1948).

85. Willson, P. D., and T. H. Edwards, *Appl. Spectrosc. Rev.*, **12**, 1 (1976).

86. Winefordner, J. D., J. J. Fitzgerald, and N. Omenetto, *Appl. Spectrosc.*, **29**, 369 (1975).

87. Yamaguchi, A., I. Ichishima, and S. Mizushima, *Spectrochim. Acta*, **12**, 294 (1958).

SPECTROSCOPIC LITERATURE

The literature of IR spectroscopy falls into two categories: research papers and books about some phase of IR spectroscopy or its use in solving problems, and compilations of spectra. In the first category, the literature is growing at such a rate (about 4000 papers per year during the first part of the 1970s, according to McDonald [102]) that one can scarcely count the papers, let alone read them. New monographs covering general and special aspects of IR spectroscopy appear frequently; a selected list is given in Table 3.1. Among the best sources of information about current developments, aside from *Chemical Abstracts,* are the biennial reviews that appear in the journal *Analytical Chemistry.* Technique-oriented fundamental reviews (e.g., IR, mass spectrometry, GC) appear with the April issue in even-numbered years; application-oriented reviews (e.g., polymers, pharmaceuticals, petroleum, rubber) appear in odd-numbered years.

INFRARED REFERENCE SPECTRA

The state of affairs in the second category is not nearly as satisfactory. Although there is no dearth of available infrared curves, they vary widely in quality and usefulness. Further, quantitative transfer of spectral information from these curves is virtually impossible, for reasons discussed earlier (see p. 51).

For qualitative analyses, good quality literature spectra can be quite useful. Some of the more readily available spectral collections are listed in Tables 3.2 and 3.3. In this author's opinion, tables of peak maxima, although better than nothing, are a good deal less useful than absorption curves because much useful information is obtained from band widths and shapes. Probably the most useful file in any laboratory is that accumulated by the user from his own instrument. Building a spectral library is discussed more fully in a subsequent section.

The importance of a good library of IR reference spectra cannot be overstressed. The wise use of correlation charts and group frequencies can give useful clues to the identity of an unknown material, but positive identification (by IR) can be made only by comparison with authenticated

TABLE 3.1 Monographs and Reviews Dealing with IR Spectroscopy

Subject	Reference
Analytical IR spectroscopy	5, 9, 72, 124
Applications	
Adsorbed species	96
Biology	81, 119, 140
Chromatographic fractions	36, 57
Coatings, paints, varnishes	24, 78
Forensics	42
Gas, vapor spectra	142, 148
Industrial spectroscopy	87, 152
Inorganics	1, 52, 62, 82, 113, 115
Isotope-labeled compounds	123
Organometallics	1, 62, 99, 114
Organophosphorus compounds	116, 141
Pharmaceuticals	10, 139
Polymers and plastics	39, 47, 70, 73, 78, 79, 158
Soils, minerals	50, 51, 58, 85, 145
Surfaces	65, 69, 89
Surfactants	77
Basic IR spectroscopy	4, 30, 33, 34, 41, 72, 86, 124, 127, 144
Bibliography of IR literature	31
Calibration of IR spectrometers	29, 128
Computer applications to IR	100, 101
Definitions	5, 44
Far IR	20, 55
Fourier transform spectroscopy	16, 53, 63, 101, 143
Gas analyzers, nondispersive	75
Group frequencies	11, 17, 18, 30, 38, 45, 83, 112
Group theory	37, 54
Hydrogen bonding	80, 122, 134
Index to literature spectra	6, 7, 8, 62
Instrumentation	
Dispersive	34, 137
Fourier transform	63, 143
Interpretation of spectra	12, 76, 135, 147
Lasers in spectroscopy	14, 107
Matrix isolation	66
Molecular spectroscopy	13, 15, 21, 74, 93, 153, 155
Optoacoustic spectroscopy	26
Physics of IR	64
Programmed learning, interpretation	12, 35, 76

TABLE 3.1 (Continued)

Subject	Reference
Series, continuing, on special topics	25, 46, 143
Techniques	
Attenuated total reflectance	68
Laboratory practice	5, 9, 47, 104, 124, 138
Reflectance	149, 157
Theoretical spectroscopy	21, 93, 136, 153
Vibration–rotation spectra	3, 74, 136, 153
Vibrational assignments	45, 133, 147

reference spectra. Reference spectra serve another useful purpose even if an exact match cannot be found; they can point the way to probable structures (or exclude postulated structures) through comparison of similar molecules containing some of the same groups. To an experienced spectroscopist, the spectrum reference library is as important as the spectrometer.

SPECTRUM-RETRIEVAL SYSTEMS

It is one thing to have a collection of spectra and quite another to be able to retrieve a particular spectrum when it is needed. Spectra can be indexed according to molecular formula, by chemical class, by absorption pattern, or by some combination of these methods. Some name and formula indexes to spectra that appear in the scientific literature are listed in Table 3.1. A formula index either to published spectra or to a private file is easy to compile and is a useful aid. Developing a scheme for identifying unknown materials from their absorption patterns is a more challenging problem.

Most spectrum retrieval systems use a binary-coding arrangement that divides the spectrum into a number of intervals of uniform width (e.g., 0.1 μm or 100 cm^{-1}). Presence of a codable absorption band in the coding interval is indicated by a "1" in the computer, by a slot in the edge of a card, or by some similar scheme. Absence of absorption is coded as "0", or no slot. The presence or absence of chemical groups or other information may be coded in the same way.

Computerized retrieval systems are often employed for matching unknown IR patterns. A number of these have been described in the literature [48, 49], and several commercial search services based on these

TABLE 3.2   General Collections of IR Spectra

American Petroleum Institute (API) Research Project 44, Chemistry Dept., Texas Agricultural and Mechanical University, College Station, Texas. About 3500 spectra, mostly of petroleum hydrocarbons and related compounds.

Coblentz Society, *Deskbook of Infrared Spectra*, C. D. Craver, Ed., P. O. Box 9952, Kirkwood, Mo. 63122. About 900 high-quality evaluated IR spectra (mostly 250–4000 cm^{-1}) illustrating the major chemical classes.

Coblentz Society Spectra (issued through Sadtler Research Laboratories). About 10,000, including 5000 evaluated spectra [27] as of 1977.

Colthup, Daly, et al. [30]. About 600 interpreted general spectra (600–4000 cm^{-1}) with reduced ordinate.

Documentation for Molecular Spectroscopy, *Working Atlas of Infrared Spectroscopy*, Butterworths, London, 1972. Contains 800 spectra covering the main groups of organic compounds (600–4000 cm^{-1}).

Infrared Data Committee of Japan, Sanyo Shuppan Boeki Co., Inc., Hoyu Bldg., 8, 2-Chome, Takara-cho, Chuo-ku, Tokyo. About 14,000 IR spectra. Most recent ones are grating spectra.

Mecke, R., and F. Langenbucher, Eds., *Infrared Spectra of Selected Chemical Compounds*, Heyden, London. Contains 1879 spectra. Wavelength data are also tabulated.

Pouchert, C. J., *The Aldrich Library of Infrared Spectra*, 2nd ed., Aldrich Chemical Co., Milwaukee, Wisc. (1975). About 11,000 spectra, arranged by chemical classes.

Sadtler Research Laboratories, *Standard Infrared Spectra*, 3314 Spring Garden St., Philadelphia, Pa. 19104. About 40,000 grating and 50,000 prism-format spectra. Special and commercial collections related to particular fields of interest are also available.

Schrader and Meier [132]. Contains IR and Raman spectra of 1014 compounds, including hydrocarbons, *N*-heterocyclics, pesticides, pharmaceuticals, polymers, and others in linear wavenumber format.

Thermodynamic Research Center Project (TRC), Chemistry Dept., Texas A. & M. University, College Station, Tex. 77843. About 1180 spectra of petrochemicals and other major industrial chemicals.

---

principles are offered. To be most useful, search systems should offer the following features:

- Ease of inputting data, using common, easily understood terms.
- No rigid requirement for an exact 1:1 correspondence of unknown and reference spectra (most real-world samples are mixtures or impure).

TABLE 3.3 Specialized Spectra Collections

Product Class	No. of Spectra	Range	Reference
Barbiturates	30	625–4000 cm^{-1}	92
	9	650–4000	98
	29	650–4000	23
Drugs, pharmaceuticals	817	2–16 μm	10
	60	250–4000	67
	175	2–15	71
	24	600–4000	90
	78	600–4000	91
	335	2–15	129
	68	50–3700	132
	268	2–15	139
	33	2–15	146
Emulsion polish components	31	2–16	111
Essential oils	200	625–4000	19
	24	300–4000	150
	36	300–4000	151
Explosives	50	200–4000	22
	176	2–15	125
	68	2–15	126
Far-IR spectra, general	1566	14–33	20
Fibers	33	2.5–15	110
Fragrance and flavor components	99	300–4000	109
Gases, vapors	66	2–16	121
	616	300–4000	142
	303	700–4000	148
Halogenated hydrocarbons	238	400–4000	28
Inorganics	875	45–3800	115
Minerals	(739)	(Tables only)	58
	?	3200–3800 and 125–700	145
Narcotics	23	2–15	97
Natural products	400	2–16	156
Paint and varnish constituents	740	250–4000	24
(Includes polymers, pigments, solvents, and additives)	2821	2.5–15	78
Pesticides	76	250–4000	59
	24	2–35	108

TABLE 3.3   (Continued)

Product Class	No. of Spectra	Range	Reference
Pigments	78	200–1500	2
Plasticizers	280	400–4000	28
	312	400–4000	78
	25	3–15	88
Polymers, plastics	1454	2–18	78
Solvents, deuterated	24	600–4000	103
Surfactants	466	2–16	77
Terpene alcohols	141	2–15	105
Terpenes	72	2–15	106
Textile finishes	84	500–4000	117
Triglycerides, purified	13	2.5–40	118

- Provision for the use of "no-band" information. Absence of a band is often more significant than its presence.
- Allowance for small wavelength errors in coding.
- Coding of known chemical features.

Desirable features include:

- Rapid file searches.
- Interaction of the searcher with the computer to permit small modifications in the input data (browsing).
- Listing of "near misses" in decreasing order of probability.

A method of quantitatively evaluating IR searching systems has been published by Erley [49], who evaluated accuracy, precision, false lookups, search time, cost, and overall figure of merit for three related searching systems. One hundred thirty-eight spectra were randomly selected from several files as "unknowns," and all 92,000 spectra in the American Society for Testing and Materials (ASTM) file were searched. The most sophisticated of the search programs gave almost as good results when used by a novice as when used by an expert.

The ASTM file contains considerable duplication of spectra; this is fortunate, since 3–5% of the entries contain errors in their redundant information [43], apart from the actual coding of the spectra. It is safe to assume at least an equal error rate in band coding. Thus some elements of forgiveness are essential in any searching routine.

All files of coded spectra that are based on the ASTM file are binary coded. Coding systems that utilize both band intensity and shape have been proposed [56, 120]. Recoding the large existing files, however, would be an arduous task, unlikely to be undertaken in the foreseeable future, although the additional data could provide some real advantages.

The question of whether a computer can be programmed to "think" in the manner of a spectroscopist interpreting a spectrum is an intriguing one. Some modest success has been achieved on a very limited basis by using a restricted compound field and additional inputs such as proton resonance and mass spectra [60, 130]. Another approach is that of pattern recognition, in which the computer uses a "learning set" of authenticated spectra to establish probable correlations that can be used to classify IR patterns according to certain chemical classes [32, 61, 94, 95, 154].

EVALUATION OF IR SPECTRA

Criteria for the evaluation of IR reference spectra have been published by the Coblentz Society [27]. Four different classes of spectra are recognized: Class I, spectra that are physical constants of a material, independent of the spectrometer on which they were taken; Class II, Research Quality Reference Spectra, spectra of pure materials measured on a good grating spectrometer operated at maximum efficiency under conditions consistent with acceptable laboratory practice; Class III, Analytical Reference Spectra, those produced on defined substances, using good laboratory techniques with a high-quality grating or prism spectrometer that does not meet Class II criteria; and finally, those spectra not classified for one reason or another. Recommendations for spectra to be published in journals, based on these criteria, have been set forth elsewhere [40] and are summarized in Table 3.4.

In any event, all spectra for whatever purpose should always have the operating conditions recorded: slit program, scan speed, spectrometer, sample form, and any other important relevant parameters.

BUILDING A PRIVATE FILE

Almost any user of an IR spectrometer will, in time, accumulate a file of reference spectra, and it is to his advantage to index and classify this material as soon as possible. In general, only spectra of pure compounds or materials of known or reproducible composition should be placed in the reference file. It is convenient to use a copy of the spectrum for reference and to keep the original spectrum in a safe location. In this connection, charts not larger than 8 in. × 24 in. allow sufficiently accurate

TABLE 3.4   Suggested Guidelines for Publication of IR Spectra:[a] Specific Criteria

Compound purity:	Minor impurity bands (no more than 3) allowed if labeled
Spectrometer resolution:	Rock-salt prism or better
Wavelength accuracy:	Within $\pm$ 5 cm^{-1}, 2000–4000 cm^{-1}; ± 3 cm^{-1} below 2000 cm^{-1}
Calibration check:	Indene or other secondary standard, run within 1 week of spectra
Noise level:	Maximum peak–peak, 1%
Energy:	At least 50% of normal in all regions (except at 4.3 μm CO_2 absorption)
Stray radiation:	Less than 5%
Extraneous absorptions:	Water absorption in pellets <0.1 absorbance unit; H_2O vapor <2% T
Servo-system energy:	Normal; deadness or excessive overshoot unacceptable
Documentation:	Specify make and model of spectrometer; prism or grating; give scan time and other relevant data, along with the physical state of sample (mull, solvent, liquid film)
Intensity:	At least one band having <25% T; no more than one band having <1% T; background line >75%; for pellets, $T > 40\%$ at 2000 cm^{-1}
Format:	Original chart, *not* hand retraced

[a] Infrared reference spectra should meet or exceed the Coblentz Society specifications established for the National Reference Data System [27], which require essentially spectra of pure compounds 650–4000 cm^{-1} with no more than minor gaps (marked) from mulling agents or solvent. Original spectra or photographic copies (*not* redrawn spectra) should be used. Contributors are encouraged to meet Class II specifications whenever possible.

readout for most purposes and are much preferable to the cumbersome charts sometimes encountered. Smaller-sized charts are more easily handled and copied; they may be stored in an ordinary file drawer and most important, a number of them can be lined up on a desk top for comparison. Reference spectra can be numbered serially or placed in groups (which, however, tend rapidly to become unwieldy). Reference spectra may be indexed by empirical formula, with a special name index for polymers, proprietary products, and materials of uncertain composition. It is important that spectra for the reference file be run in a standard, reproducible manner.

For the worker in a small- or medium-sized laboratory who wishes to code his own reference file of IR spectra, several options are available. Edge-punched, hand-sorted cards can be used if the file does not exceed about 2000 cards. The optical coincidence or "peak-a-boo" system [131] can accommodate up to 10,000 compounds on 250 cards. Some commercial computerized search services allow addition of a private file, suitably protected so that it can be used only by its owner.

When coding spectra, one should not code every band, since over-punching removes the uniqueness of the pattern just as underpunching does. The criterion used in the ASTM file is that all bands should be coded that have an absorbance ratio greater than 1 : 10 with the strongest band in the spectrum. Other workers use a 1 : 5 ratio so that even fewer bands are punched, which is an advantage for negative (no band) sorting procedures. In the latter case, 5–8 bands per compound will be coded.

References

1. Adams, D. M., *Metal–Ligand and Related Vibrations,* St. Martin's Press, New York, 1968.

2. Afremow, L. C., and J. T. Vandeberg, *J. Paint Technol.,* **38,** 169 (1966).

3. Allen, H. C., Jr., and P. C. Cross, *Molecular Vib-Rotors,* Wiley, New York, 1963.

4. Alpert, N. L., W. E. Keiser, and H. A. Szymanski, *IR; Theory and Practice of Infrared Spectroscopy,* 2nd ed., Plenum Press, New York, 1970.

5. American Society for Testing and Materials, *1977 Annual Book of ASTM Standards, Part 42,* ASTM, Philadelphia, 1977.

6. American Society for Testing and Materials, *Alphabetical List of Compound Names, Formulas, and References to Published Infrared Spectra; an Index to 92,000 Published Infrared Spectra.* ASTM, Philadelphia, 1969.

7. American Society for Testing and Materials, *Molecular Formula List of Compounds, Names, and References to Published Infrared Spectra,* AMD 31, ASTM, Philadelphia, 1969.

8. American Society for Testing and Materials, *ASTM Serial Number List of Compound Names and References to Published Infrared Spectra,* Vol. 3, ASTM, Philadelphia, 1969.

9. American Society for Testing and Materials, Committee E-13, *Manual on Recommended Practices in Spectrophotometry,* ASTM, Philadelphia, 1966.

10. Association of Official Analytical Chemists, *Infrared and Ultraviolet Spectra of Some Compounds of Pharmaceutical Interest,* revised ed., AOAC, Washington, D.C., 1972.

11. Avram, M., G. D. Mateescu, *Infrared Spectroscopy: Applications in Organic Chemistry,* Wiley Interscience, New York, 1972.

12. Baker, A. J., T. Cairns, G. Eglinton, and F. J. Preston, *More Spectroscopic Problems in Organic Chemistry*, 2nd ed., Heyden, London, 1975.

13. Banwell, C. N., *Fundamentals of Molecular Spectroscopy*, 2nd ed., Mc-Graw Hill, New York, 1973.

14. Barnes, A. J., and W. J. Orville-Thomas, *Vibrational Spectroscopy—Modern Trends*, Elsevier, New York, 1977.

15. Barrow, G. M., *Introduction to Molecular Spectroscopy*, McGraw-Hill, New York, 1962.

16. Bell, R. J., *Introductory Fourier Transform Spectroscopy*, Academic Press, New York, 1972.

17. Bellamy, L. J., *Advances in Infrared Group Frequencies*, Barnes and Noble, New York, 1968.

18. Bellamy, L. J., *The Infrared Spectra of Complex Molecules*, Vol. 1, 3rd ed., Wiley, New York, 1975.

19. Bellanato, J., and A. H. Hidalgo, *Infrared Analysis of Essential Oils*, Heyden, London, 1971.

20. Bentley, F. F., L. D. Smithson, and A. L. Rozek, *Infrared Spectra and Characteristic Frequencies 700–300 cm^{-1}*, Interscience, New York, 1968.

21. Califano, S., *Vibrational States*, Wiley, New York, 1976.

22. Chasan, D. E., and G. Norwitz, *Microchem. J.*, **17**, 31 (1972).

23. Chatten, L. G., and L. Levi, *Appl. Spectrosc.*, **11**, 177 (1957).

24. Chicago Society of Paint Technology, *Infrared Spectroscopy: Its Use in the Coatings Industry*, Federat. Soc. Paint Technology, Philadelphia, 1969.

25. Clark, R. J. H., and R. E. Hester, *Advances in Infrared and Raman Spectroscopy*, Vol. 1, Heyden, London, 1975; Vol. 2, 1976; Vol. 3, 1977; Vol. 4, 1978.

26. Claspy, D. C., Infrared Optoacoustic Spectroscopy and Detection, in Y.-H. Pao, Ed., *Optoacoustic Spectroscopy and Detection*, Academic Press, New York, 1977.

27. Coblentz Society Board of Managers, *Anal. Chem.*, **38**, (9), 27A (1966); **47**, 945A (1975).

28. Coblentz Society Inc., Special Collections: CSC-2, Halogenated Hydrocarbons; CSC-3, Plastizers and Other Additives (1977).

29. Cole, A. R. H., *Tables of Wavenumbers for the Calibration of Infrared Spectrometers*, 2nd ed., Pergamon Press, Oxford, 1977.

30. Colthup, N. B., L. H. Daly, and S. E. Wiberley, *Introduction to Infrared and Raman Spectroscopy*, 2nd ed., Academic Press, New York, 1975.

31. Comeford, J. J., C. N. R. Rao, S. K. Dikshit, S. A. Kudchadkev, and D. S. Gupta, *Bibliography of Infrared Spectroscopy Through 1960*, National Bureau of Standards, Institute of Applied Technology, Washington, D.C., 1976, Paper No. NBS-SP-428, Parts 1–3.

32. Comerford, J. M., P. G. Anderson, W. H. Snyder, and H. S. Kimmel, *Spectrochim. Acta*, **33A**, 651 (1977).

33. Conley, R. T., *Infrared Spectroscopy*, 2nd ed., Allyn and Bacon, Boston, 1972.

34. Conn, G. K. T., and D. G. Avery, *Infrared Methods*, Academic Press, New York, 1960.

35. Cook, B. W., and K. Jones, *A Programmed Introduction to Infrared Spectroscopy*, Heyden, London, 1972.

36. Copier, H., *Infrared Analysis of Chromatographic Fractions*, Druk. Elinkwijk, Utrecht, The Netherlands, 1968.

37. Cotton, F. A., *Chemical Applications of Group Theory*, Interscience, New York, 1963.

38. Craver, C. D., *Desk Book of Infrared Spectra*, Coblentz Society, POB 9952, Kirkwood, Mo., 1977.

39. Craver, C. D., *Polymer Characterization: Interdisciplinary Approaches*, Plenum Press, New York 1971.

40. Craver, C. D., J. G. Grasselli, and A. L. Smith, *Anal. Chem.*, **47**, 2065 (1975).

41. Cross, A. D., *Introduction to Practical Infrared Spectroscopy*, 2nd ed., Butterworth, London, 1964.

42. De Faubert Maunder, M. J., *Practical Hints on Infrared Spectrometry for a Forensic Analyst*, Hilger, London, 1971.

43. De Haseth, J. A., H. B. Woodruff, and T. L. Isenhour, *Appl. Spectrosc.*, **31**, 18 (1977).

44. Denney, R. C., *A Dictionary of Spectroscopy*, Halsted Press, New York, 1973.

45. Dolphin, D., and A. E. Wick, *Tabulation of Infrared Spectral Data*, Wiley, New York, 1977.

46. Durig, J. R., *Vibrational Spectra and Structure, A Series of Advances*, Vol. 4, Elsevier, Amsterdam, 1975; Vol. 5, 1976; Vol. 6, 1977.

47. Elliott, A., *Infrared Spectra and Structure of Organic Long-Chain Polymers*, Arnold, London, 1969.

48. Erley, D. S., *Anal. Chem.*, **40**, 894 (1968).

49. Erley, D. S., *Appl. Spectrosc.*, **25**, 200 (1971).

50. Farmer, V. C., Ed., *The Infrared Spectra of Minerals*, Mineralogical Society, London, 1974.

51. Farmer, V. C., and F. Palmieri, Characterization of Soil Minerals by IR Spectroscopy, in J. E. Gieseking, Ed., *Soil Components*, Vol. 2, Springer, New York, 1975, p. 573.

52. Ferraro, J. R., *Low-Frequency Vibrations of Inorganic and Coordination Compounds*, Plenum Press, New York, 1971.

53. Ferraro, J. R., and L. J. Basile, *Fourier Transform Infrared Spectroscopy*, Academic Press, New York, 1977.

54. Ferraro, J. R., and J. S. Ziomek, *Introductory Group Theory and Its Application to Molecular Structure*, 2nd ed., Plenum Press, New York, 1975.

55. Finch, A., P. N. Gates, K. Radcliffe, F. N. Dickson, and F. F. Bentley, *Chemical Applications of Far Infrared Spectroscopy,* Academic Press, New York (1970).

56. Fox, R. C., *Anal. Chem.,* **48,** 717 (1976).

57. Freeman, S. K., Gas Chromatography and Infrared and Raman Spectrometry, in L. S. Ettre and W. H. McFadden, Eds., *Ancillary Techniques of Gas Chromatography,* Wiley, New York, 1969, pp. 227–267.

58. Gadsden, J. A., *Infrared Spectra of Minerals and Related Inorganic Compounds,* Butterworth, London, 1975.

59. Gore, R. C., R. W. Hannah, S. C. Pattacini, and T. J. Porro, *J. Assoc. Offic. Anal. Chem.,* **54,** 1040 (1971).

60. Gray, N. A. B., *Anal. Chem.,* **47,** 2426 (1975).

61. Gray, N. A. B., *Anal. Chem.,* **48,** 2265 (1976).

62. Greenwood, N. N., E. J. F. Ross, and B. P. Straughan, *Index of Vibrational Spectra of Inorganic and Organometallic Compounds,* Vol. 1, Butterworth, London, 1972, and succeeding volumes.

63. Griffiths, P. R., *Chemical Infrared Fourier Transform Spectroscopy,* Wiley, New York, 1975.

64. Hackforth, H. C., *Infrared Radiation,* McGraw-Hill, New York, 1960.

65. Hair, M. L., *Infrared Spectroscopy in Surface Chemistry,* Marcel Dekker, New York, 1967.

66. Hallam, H. E., *Vibrational Spectroscopy of Trapped Species,* Wiley, New York, 1973.

67. Hannah, R. W., and S. C. Pattacini, *The Identification of Drugs from Their Infrared Spectra,* Perkin-Elmer Applications Study No. 11, 1972.

68. Harrick, N. J., *Internal Reflection Spectroscopy,* Interscience, New York, 1967.

69. Harrick, N. J., and K. H. Beckmann, in P. F. Kane, Ed., *Characterization of Solid Surfaces,* Plenum Press, New York, 1974, pp. 215–245.

70. Haslam, J., and H. A. Willis, *Identification and Analysis of Plastics,* Van Nostrand, Princeton, N. J., 1965.

71. Hayden, A. L., O. R. Sammul, G. B. Selzer, and J. Carol, *J. Assoc. Offic. Agric. Chem.,* **45,** 797 (1962).

72. Hediger, H., *Infrared Spectroscopy: Principles, Uses, Interpretation,* Vol. 11, *Methods of Analysis in Chemistry,* Akad. Verlagsges, Frankfurt, 1971.

73. Henniker, J. C., *Infrared Spectrometry of Industrial Polymers,* Academic Press, New York, 1967.

74. Herzberg, G., *Molecular Spectra and Molecular Structure. II. Infrared and Raman Spectra of Polyatomic Molecules,* Van Nostrand, New York, 1945.

75. Hill, D. W., and T. Powell, *Nondispersive Infrared Gas Analysis in Science, Medicine and Industry,* Hilger, London, 1968.

76. Hill, R. R., and D. A. E. Rendell, *The Interpretation of Infrared Spectra,* Heyden, London, 1975.

77. Hummel, D. O., *Identification and Analysis of Surface-Active Agents,* Wiley-Interscience, New York, 1962.

78. Hummel, D. O., and F. K. Scholl, *Infrared Analysis of Polymers, Resins, and Additives: an Atlas,* Vol. 1, *Plastics, Elastomers, Fibers, and Resins,* Part 1: Text; Part 2, Spectra, Tables, Index, Wiley-Interscience, New York, 1969; Vol. 2, *Additives and Processing Aids,* 1973.

79. Ivin, K. J., *Structural Studies of Macromolecules by Spectroscopic Methods,* Wiley, New York, 1976.

80. Joesten, M. D., and L. J. Schaad, *Hydrogen Bonding,* Marcel Dekker, New York, 1974.

81. Jones, D. W., *Introduction to the Spectroscopy of Biological Polymers,* Academic Press, New York, 1976.

82. Jones, L. H., *Inorganic Vibrational Spectroscopy,* Vol. 1, Marcel Dekker, New York, 1971.

83. Jones, R. N., *Infrared Spectra of Organic Compounds: Summary Charts of Principal Group Frequencies,* National Research Council of Canada, Ottawa, 1959.

84. Jurs, P. C., and T. L. Isenhour, *Chemical Applications of Pattern Recognition,* Wiley-Interscience, New York, 1975.

85. Karr, C., Jr., *Infrared and Raman Spectroscopy of Lunar and Terrestrial Minerals,* Academic Press, New York, 1975.

86. Kemmner, G., *Infrared Spectroscopy: Principles, Applications, Methods,* Vol. 5, *Chemical Monographs,* Franckh, Stuttgart, 1969.

87. Kendall, D. N., *Applied Infrared Spectroscopy,* Reinhold, New York, 1966.

88. Kendall, D. N., R. R. Hampton, H. Hausdorf, and F. Pristera, *Appl. Spectrosc.,* **7,** 179 (1953).

89. Kiselev, A. V., and V. I. Lygin, *Infrared Spectra of Surface Compounds,* Halsted Press, New York, 1975.

90. Lenzen, C., and L. Delcambe, *Inf. Bull., Int. Cent. Inf. Antibiot.,* **10,** 78 (1972).

91. Lenzen, C., and L. Delcambe, *Inf. Bull., Int. Cent. Inf. Antibiot.,* **11,** 157 (1973).

92. Levi, L., and C. E. Hubley, *Anal. Chem.,* **28,** 1591 (1956).

93. Levin, I. N., *Molecular Spectroscopy,* Wiley, New York, 1975.

94. Liddell, R. W., III, and P. C. Jurs, *Anal. Chem.,* **46,** 2126 (1974).

95. Liddell, R. W., and P. C. Jurs, *Appl. Spectrosc.,* **27,** 371 (1973).

96. Little, L. H., *Infrared Spectra of Adsorbed Species,* Academic Press, New York, 1966.

97. Manning, J. J., *Appl. Spectrosc.,* **10,** 85 (1956).

98. Manson, J. M., and J. A. R. Cloutier, *Appl. Spectrosc.,* **15,** 77 (1961).

99. Maslowsky, E., Jr., *Vibrational Spectra of Organometallic Compounds,* Wiley, New York, 1976.

100. Mattson, J. S., H. B. Mark, Jr., and H. C. MacDonald, Jr., *Computers in Chemistry and Instrumentation*, Vol. 5, *Laboratory Systems and Spectroscopy*, Marcel Dekker, New York, 1977.

101. Mattson, J. S., H. B. Mark, Jr., and H. C. MacDonald, Jr., *Computers in Chemistry and Instrumentation*, Vol. 7, *Infrared, Correlation, and Fourier Transform Spectroscopy*, Marcel Dekker, New York, 1977.

102. McDonald, R. S., *Anal. Chem.*, **48**, 196R (1976).

103. McNiven, N. L., and R. Court, *Appl. Spectrosc.*, **24**, 296 (1970).

104. Miller, R. G. J., and B. C. Stace, *Laboratory Methods in Infrared Spectroscopy*, 2nd ed., Heyden, London, 1972.

105. Mitzner, B. M., V. J. Mancini, S. Lemberg, and E. T. Theimer, *Appl. Spectrosc.*, **22**, 34 (1968).

106. Mitzner, B. M., E. T. Theimer, and S. K. Freeman, *Appl. Spectrosc.*, **19**, 169 (1965).

107. Mooradian, A., T. Jaeger, and P. Stokseth, *Tunable Lasers and Applications*, Vol. 3, Springer Series in Optical Sciences, Springer-Verlag, Berlin, 1976.

108. Morris, W. M., Jr., and E. O. Haenni, *J. Assoc. Offic. Agric. Chem.*, **46**, 964 (1963).

109. Morris, W. W., *J. Assoc. Offic. Anal. Chem.*, **56**, 1037 (1973).

110. Morrison, R. D., *Am. Dyest. Rep.*, **52**, 867 (1963); ASTM Test D 276-72, *1976 Book of ASTM Standards*, Part 33, Textiles—Fibers and Zippers; High Modulus Fibers, American Society for Testing and Materials, Philadelphia, 1976, p. 46.

111. Murphy, J. E., and W. C. Schwemer, *Anal. Chem.*, **30**, 116 (1958).

112. Nakanishi, K., and P. H. Solomon, *Infrared Absorption Spectroscopy*, 2nd ed., Holden-Day, San Francisco, 1977.

113. Nakomoto, K., *Infrared Spectra of Inorganic and Coordination Compounds*, 2nd ed., Wiley-Interscience, New York, 1970.

114. Nakamoto, K., and P. J. McCarthy, *Spectroscopy and Structure of Metal Chelate Compounds*, Wiley, New York, 1968.

115. Nyquist, R. A., and R. O. Kagel, *Infrared Spectra of Inorganic Compounds*, Academic Press, New York, 1971.

116. Nyquist, R. A., and W. J. Potts, Jr., Vibrational Spectra of Phosphorus Compounds, in M. Halman, Ed., *Analytical Chemistry of Phosphorus Compounds*, Interscience, New York, 1972.

117. O'Connor, R. T., E. R. McCall, N. M. Morris, and V. W. Tripp, *U.S. Agric. Res. Serv., South. Reg. [Rep.]*, report No. ARS-S-47, 1974.

118. Parkash, S., and J. M. V. Blanshard, *Spectrochim. Acta*, **31A**, 951 (1975).

119. Parker, F. S., *Applications of Infrared Spectroscopy in Biochemistry, Biology, and Medicine*, Plenum Press, New York, 1971.

120. Penski, E. C., D. A. Padowski, and J. B. Bouck, *Anal. Chem.*, **46**, 955 (1974).

121. Pierson, R. H., A. N. Fletcher, and E. S. Gantz, *Anal. Chem.*, **28**, 1218 (1956).

122. Pimentel, G. C., and A. L. McClellan, *The Hydrogen Bond,* Freeman, San Francisco, 1960.

123. Pinchas, S., and I. Laulicht, *Infrared Spectra of Labeled Compounds,* Academic Press, New York, 1971.

124. Potts, W. J., Jr., *Chemical Infrared Spectroscopy,* Vol. 1, *Techniques,* Wiley, New York, 1963.

125. Pristera, F., and W. E. Fredericks, *U.S. Clearinghouse Fed. Sci. Technol. Inform.,* AD 1969, No. 859846.

126. Pristera, F., M. Halik, A. Castelli, and W. Fredericks, *Anal. Chem.*, **32**, 495 (1960).

127. Rao, C. N. R., *Chemical Applications of Infrared Spectroscopy,* Academic Press, New York, 1963.

128. Rao, K. N., C. J. Humphreys, and D. H. Rank, *Wavelength Standards in the Infrared,* Academic Press, New York, 1966.

129. Sammul, O. R., W. L. Brannon, and A. L. Hayden, *J. Assoc. Offic. Agric. Chem.*, **47**, 918 (1964).

130. Sasaki, S., H. Abe, T. Ouki, M. Sakamoto, and S. Ochiai, *Anal. Chem.*, **40**, 2220 (1968).

131. Schlichter, N. E., and E. Wallace, *Appl. Spectrosc.*, **17**, 98 (1963).

132. Schrader, B., and W. Meier, *Raman/IR Atlas,* Vol. 1, Verlag Chemie, Weinheim, Germany, 1974.

133. Shimanouchi, T., *Tables of Molecular Vibrational Frequencies, Consolidated Volume I.* U.S. Government Printing Office, S. D. Cat. No. C13.48:39, 1972.

134. Shuster, P., G. Zundel, and C. Sandorfy, *The Hydrogen Bond: Recent Developments in Theory and Experiments,* Vol. 2, *Structure and Spectroscopy,* North-Holland, Amsterdam, 1976.

135. Silverstein, R. M., G. C. Bassler, and T. C. Morrill, *Spectrometric Identification of Organic Compounds,* 3rd ed., Wiley, New York, 1974.

136. Steele, D., *Theory of Vibrational Spectroscopy,* Saunders, Philadelphia, 1971.

137. Stewart, J. E., *Infrared Spectroscopy: Experimental Methods and Techniques,* Marcel Dekker, New York, 1970.

138. Stine, K. E., *Modern Practices in Infrared Spectroscopy,* Beckman Instruments, Fullerton, Ca., 1970.

139. Sunshine, I., and S. R. Gerber, *Spectrophotometric Analysis of Drugs,* Charles C. Thomas, Springfield, Ill., 1963.

140. Thomas, G. J., Jr., and Y. Kyogoku, Biological Science, in E. G. Brame, Jr., and J. G. Grasselli, Eds., *Practical Spectroscopy,* Vol. 1, *Infrared and Raman Spectroscopy,* Marcel Dekker, New York, 1977, Part C, p. 717.

141. Thomas, L. C., *Interpretation of the Infrared Spectra of Organophosphorus Compounds*, Heyden, London, 1974.

142. Thompson, B., *Hazardous Gases and Vapors: Infrared Spectra and Physical Constants*, Beckman Instruments, Fullerton, Ca., technical report No. 595, August 1974.

143. Vanasse, G. A., *Spectrometric Techniques*, Vol. 1, Academic Press, New York, 1977.

144. Van der Maas, J. H., *Basic Infrared Spectroscopy*, 2nd ed., Heyden, London, 1972.

145. Van der Marel, H. W., and H. Beutelspacher, *Atlas of Infrared Spectroscopy of Clay Minerals and Their Admixtures*, Elsevier, New York, 1976.

146. Wayland, L. and P. J. Weiss, *J. Assoc. Offic. Agric. Chem.*, **48**, 965 (1965).

147. Weitkamp, H., and R. Barth, *Infrared Structural Analysis: a Dualistic Interpretation Scheme*, Thieme, Stuttgart, Germany, 1972.

148. Welti, D., *Infrared Vapour Spectra*, Heyden, London, 1970.

149. Wendlandt, W. W., and H. G. Hecht, *Reflectance Spectroscopy*, Interscience, New York, 1966.

150. Wenninger, J. A., and R. L. Yates, *J. Assoc. Offic. Anal. Chem.*, **53**, 949 (1970).

151. Wenninger, J. A., R. L. Yates, and M. Dolinski, *J. Assoc. Offic. Anal. Chem.*, **50**, 1313 (1967).

152. White, R. G., *Handbook of Industrial Infrared Analysis*, Plenum Press, New York, 1964.

153. Wilson, E. B., Jr., J. C. Decius, and P. C. Cross, *Molecular Vibrations*, McGraw-Hill, New York, 1955.

154. Woodruff, H. B., G. L. Ritter, S. R. Lowry, and T. L. Isenhour, *Appl. Spectrosc.*, **30**, 213 (1976).

155. Woodward, L. A., *Introduction to the Theory of Molecular Vibrations and Vibrational Spectroscopy*, Oxford U.P., London, 1972.

156. Yamaguchi, K., *Spectral Data of Natural Products*, Vols. 1 and 2, Elsevier, New York, 1970.

157. Young, E. F., and R. W. Hannah, in W. W. Wendlandt, Ed., *Modern Aspects of Reflectance Spectroscopy*, Plenum Press, New York, 1968.

158. Zbinden, R., *Infrared Spectroscopy of High Polymers*, Academic Press, New York, 1964.

SAMPLING TECHNIQUES

An almost infinite variety of sampling techniques can be used to obtain an IR spectrum, and the analyst must choose the one that best fits his problem. Some of the most useful methods are described here; however, many variations of these methods are possible. A manual of recommended practices that covers sample preparation, qualitative and quantitative analyses, and other topics is available from the ASTM [3].

At times the sampling method is dictated by the nature of the sample, but usually some choice is available. It is a strong temptation for the neophyte spectroscopist to run all liquids as smears between salt plates and all solids as KBr pellets. Although these techniques each have a place, they have serious limitations and most certainly should not be adopted as standard procedures for all samples.

Because the physical state of a sample may have a profound effect on its IR spectrum, it is wise to determine in advance a hierarchy of sampling methods that will be used within the laboratory. This hierarchy is determined by the types of samples encountered and the sampling methods used for spectra in the reference library. In a laboratory doing general chemical work, for example, the preferred sequence for liquids might be: (1) solution, (2) if insoluble, undiluted in a thin cell, and (3) liquid film between salt plates. For powders and friable solids, a logical sampling order might be: (1) solution, (2) mineral-oil mull, (3) KBr pellet, and (4) pyrolysate. Techniques such as attenuated total reflectance (ATR) are generally reserved for special sampling situations.

SOLUTIONS

In many situations, sampling in solution is preferred. This technique, although slightly more troublesome than some others, has the tremendous advantage of being exactly reproducible. The spectral record thus can be compared with spectra taken perhaps years later, and minor points of difference can be easily recognized. Semiquantitative analysis can be done easily if concentration and cell thickness are recorded (as they should be). Furthermore, concentrations may be matched to cell thickness in such a way that the shape and structure of all the strong bands are clearly evident, a situation that does not often occur when liquids are run undiluted in fixed-thickness cells. (The exceptions to this statement are aliphatic hydrocarbons and aliphatic solvents, which give more revealing spectra if they are run undiluted.) Finally, in the case of solids, the effects of polymorphism, which can seriously influence the spectral pattern, are eliminated.

Solvents

Choice of solvent is always a compromise. Since all common solvents absorb in the IR region, one must use thin layers and choose solvents that have windows in the regions of interest. It is not always easy to find a transparent solvent in which the sample is completely soluble at usable concentration levels. (Occasionally, however, analysis of mixtures may be considerably simplified by taking advantage of selective extraction by the IR solvent.)

The solvent should be chemically inert to the sample. Primary and secondary amines, for example, react with CS_2 to give thiocarbamates and are best sampled as liquid films. Aliphatic amines also undergo a slow photochemical reaction with CCl_4 that forms the amine hydrochloride. Moisture-sensitive materials may react with residual water in any solvent, especially at large dilutions, unless the solvent is carefully dried.

On first impression, it might appear that any solvent can be used with double-beam spectrometers, since the absorption in the sample beam is cancelled by solvent in the reference beam. In regions of strong solvent absorption, however, no energy passes into the spectrometer and the instrument is said to be "dead."

For general work in the $625-4000$-cm^{-1} ($2.5-16$-μm) region, it is common practice to use CCl_4 or C_2Cl_4 at $1330-4000$ cm^{-1} ($2.5-7.5$ μm) and CS_2 at $625-1330$ cm^{-1} ($7.5-16$ μm). No solvent is completely without influence on the solute, but nonpolar solvents such as these show a minimum effect because of their relatively homogeneous dielectric field. It should be noted that vapors of CCl_4 and CS_2 are extremely toxic. The maximum average atmospheric concentration which a worker may be exposed without injury to health is 10 ppm for CCl_4 and 20 ppm for CS_2. Solvent odor may not be a reliable warning, since the odor threshold for CCl_4 is around 50 ppm. The olfactory senses can detect about 1 ppm of CS_2 in air, but the odor-detection limit increases on exposure. In addition, CS_2 is flammable, so manipulations of these solvents should be undertaken only in a well-ventilated hood.

For a solvent pair of reduced toxicity (but more background absorption), one may use C_2Cl_4 at 1000–4000 cm^{-1} (2.5–10 μm) and n-heptane at 250–1000 cm^{-1} (10–40 μm).

Other solvents useful over more limited ranges include $HCCl_3$, dioxane, and dimethyl formamide. The solvent action of such materials is due to their polar nature, which implies strong IR absorption as well as interaction with the solute. Deuterated solvents have "windows" that differ from those of the protonated species, and are not excessively costly in the small quantities needed [83].

A greater choice of solvents is available for use at longer wavelengths. For the 260–625-cm^{-1} (16–38-μm) region, CCl_4, dichloromethane, and hexane are useful. Many other solvents have only one or two bands and may be used over limited spans.

Solvents and window materials should be matched in refractive index to minimize interference fringes that may appear in the background (see pp. 236–239).

Pure solvent in a matching cell is usually used in the reference beam to cancel out minor background fluctuations, but it must always be kept in mind that where solvent transmission drops to less than about 30%, the servo system will not function properly and the recorder trace will be inaccurate unless compensating adjustments in gain are made.

In special situations, H_2O and/or D_2O can be used as solvents. By using very thin cells (ca. 0.02 mm) of KRS-5, BaF_2, AgCl, or Irtran-2, and by adjusting the spectrometer slits to compensate for the energy loss, one can obtain quite acceptable spectra of water solutions in the 830–1540-cm^{-1} (6.5–12-μm) range [20, 46, 100]. A different spectral range is available with D_2O, but if this material is used with samples having labile hydrogen, exchange will occur. Water solutions can also be sampled by ATR.

Moisture in samples will not only affect their solubility, but may also fog cell windows. Wet material may be dried before dissolving by adding 2,2-dimethoxy propane [30], which reacts with water in acid medium to form methanol and acetone. Both the reactant and the products are volatile and may be removed by warming.

Concentration

Solutions are usually prepared by weighing a specified amount of sample into a volumetric flask and then filling to the mark with solvent. Most organic materials give good spectra at a concentration of 1 g/10 cm^3 in a 0.1-mm cell in the 625–4000-cm^{-1} (2.5–16-μm) region. Strong absorbers such as organofluorine and organosilicon compounds are diluted to 0.2

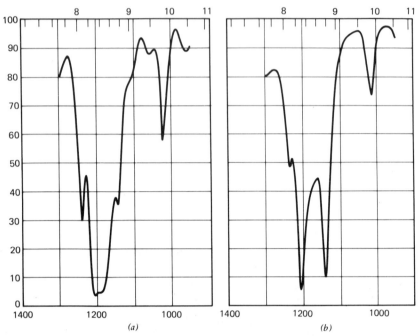

Fig. 4.1 Effect of dilution on spectrum of *t*-butyl alcohol: (*a*) 10% CS$_2$ solution in a 0.1-mm cell; (*b*) 1% solution in a 1-mm cell.

g/10 cm^3 for the 625–1330-cm^{-1} (7.5–16-μm) range. Greater concentrations or thicknesses may be needed below 600 cm^{-1} except for very strong absorbers.

Reactive or volatile materials may be sampled by weighing them into a stoppered flask half full of solvent. Even material boiling below room temperature can be handled in this way if it is first cooled below its boiling point. Special equipment can be used to contain volatile, reactive, or radioactive solutions [87].

It is important to standardize on both solvents and dilution factors, since spectra may change strikingly on dilution or on changing solvents [Fig. 4.1].

Sample Thickness

Choice of cell thickness may be influenced by the amount of sample available or by its solubility. Extremely thin cells (<0.05 mm) are difficult not only to construct, but also to fill and empty, whereas cells thicker than 0.2 mm may lead to excessive solvent absorption. A thickness of 0.1 mm is a convenient compromise. For trace analysis, cells up to 5 cm in length can be used in regions of high solvent transparency. Satisfactory transmission in the region of interest should be verified before thick section sampling is undertaken, however.

Cell Blanks

It is to be expected that the cell windows will eventually become fogged or accumulate absorbing deposits. Since the condition of a window is not always evident from its appearance, it is wise to monitor the cell periodically by running a compensated solvent blank. A daily cell match spectrum can save the analyst many fruitless hours of trying to identify spurious bands.

FILMS

Among the simpler methods of sampling is the smear or liquid-film technique. It is applicable to nonvolatile, nonreactive samples and is useful for insoluble liquids or for qualitative survey spectra. A drop of sample is squeezed between two salt plates or put onto a flat glass surface and "wiped" with a single plate. It is desirable to obtain a uniform thickness over the area intercepted by the sample beam of the spectrometer. Obviously, spectra obtained in this way are not very reproducible, and some trial and error may be required before a usable spectrum is obtained.

The smear method is also useful for sampling resins, varnishes, and other materials that are dissolved in a volatile solvent. The thin layer of resin dries very rapidly under a heat lamp and spectra are usually free from solvent interference.

Water emulsions are often sampled by drying a few drops on an AgCl window. Emulsions may be sampled directly by use of ATR methods.

Soluble polymers also may be cast on glass, smooth plastic, or mercury and run as unsupported films. Such samples often show interference fringes in transmission, which confuse the spectral pattern. If the fringes are a problem, they may be avoided by using ATR techniques. Alternatively, the film may be tilted so the radiation beam strikes it at Brewster's angle [58].

MULLS

The problem of sampling solids that are insoluble in the usual IR-transmitting solvents is most often met by preparing a mineral oil or KBr mull of the ground powder. In both cases the objectives are to provide a uniform distribution of particles in the beam and to improve transmission by suspending the particles in a medium whose refractive index is close to that of the sample.

Mineral-oil Mulls

Mineral oil is used extensively as a mulling agent but has the disadvantage of strong absorption in the CH stretching and bending regions. This objection can be overcome by using a "split mull," in which chlorinated or fluorinated oils are used for the $1340-4000$-cm^{-1} ($2.5-7.5$-μm) range and mineral oil for frequencies lower than 1340 cm^{-1} (7.5 μm).

Good mulls are not difficult to prepare if the proper technique is used [14]. It is essential that the particles be ground until they are smaller than the wavelength of the radiation used. To accomplish this objective, a small amount of the solid (the less the better; not over 10–20 mg) is first ground gently in a *large* agate mortar, and the particles are then rubbed to a fine powder by vigorous action with the pestle. After the powder has formed a glossy cake in the mortar, one drop of mineral oil is added and the grinding action is continued until the solid is completely suspended in oil. The translucent paste may then be transferred to a salt window with the help of a rubber policeman and a second window pressed on top of the paste.

The secret to preparation of good mulls is thorough grinding of the solid. With small samples, this step can ordinarily be accomplished in 1–5 min. The effects of good and poor grinding are seen in Fig. 4.2. When a solid is incompletely ground, some portions of the sample beam are intercepted by essentially opaque particles, others by particles of the proper size, and still others by no sample at all. The radiation that is not absorbed at an absorbing wavelength will then give a false zero level and the result will be a distorted spectrum. Another distortion, noted in Fig. 4.3, is due to the Christiansen effect [101], which is caused by large changes in the refractive index of the sample in the vicinity of an absorption band (cf. Fig. 4.9). Since the amount of scattered radiation is proportional to the square of the difference in refractive indices between the sample and the matrix, the sample becomes highly transmitting where the indices match on one side of the absorption band and highly scattering on the other. The resulting asymmetric absorption may resemble a derivative curve rather than an absorption spectrum. The effect is minimized by reducing the sample particles to fineness less than the wavelength of incident radiation. Mechanical grinding in a small ball mill or vibrator-grinder is also effective, but it is not as easy to follow the progress of the grinding as with the hand mortar and pestle. If the crystal structure is changed by the heat and pressure of mechanical grinding (as it frequently is), the spectrum will change as a function of grinding time.

Potassium Bromide Mulls (Pellets)

The KBr mull or pressed-pellet technique, first introduced in 1952 [105, 111], involves mixing a finely divided sample intimately with KBr (or other alkali halide) powder and then pressing the powder in a die to form a transparent or translucent pellet. Better results are usually obtained if the die is evacuated to eliminate occluded air from the pellet. Advantages of the pellet method are: (1) freedom from most interfering absorptions, (2) good control of sample concentration, and (3) convenient storage of specimens. There are also significant disadvantages, which are discussed later.

The same requirements on particle size exist for the successful preparation of KBr pellets as for mulls, and the same grinding technique may be used for the sample. Best results are obtained by thoroughly grinding the sample (alone), and then mixing (*not* grinding) the powdered sample with the powdered KBr. Hand grinding of the sample in an agate mortar is preferred. At times, sample dispersal can be expedited by adding a few drops of a suitable solvent such as methylene chloride or hexane to the KBr-sample mixture. The solvent is subsequently evaporated while light grinding is continued.

Optimum dispersal of soluble samples in the KBr matrix can be achieved by freeze drying. This technique is particularly applicable to water-soluble samples but can be used also with materials soluble in organic solvents [101, 106]. An aqueous solution of the sample and KBr is frozen as a thin film of ice on the sides of an evacuable glass container using dry ice–acetone or liquid air. The vessel is then evacuated for a period of several hours, while water sublimes, leaving a dry, fluffy powder. The powder is pressed into a pellet using standard or micro techniques, depending on the sample size. Water-insoluble materials are prepared as solutions in an organic solvent whose sublimation properties are similar to those of water. The solution is frozen inside a shell of frozen KBr solution. Lyophilization is then carried out as usual.

The KBr pellet method has been found particularly valuable in the study of cellulose [90, 91] and wool [113] fibers, in which extensive grinding is likely to introduce changes in crystallinity. The refractive index of KBr matches that of these fibers quite closely, so satisfactory spectra can be obtained from relatively coarse particles.

An ingenious method for analyzing surfaces of paints, varnishes, plas-

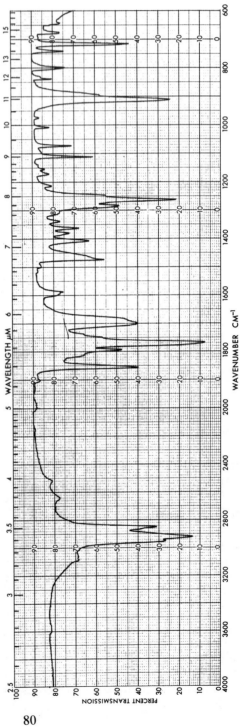

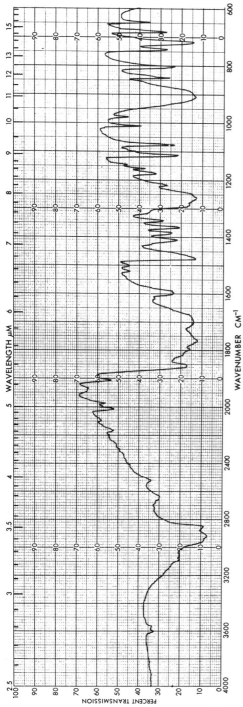

Fig. 4.2 Phthalic anhydride prepared as mineral-oil mull: *top*, improperly ground; *bottom*, properly ground.

81

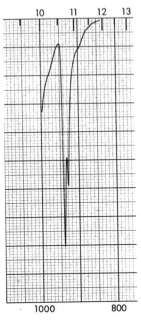

Fig. 4.3 Christiansen effect in mercuric acetate (mineral-oil mull).

tics, metals, or glass that utilizes the KBr pellet method has been described by Johnson [69, 70]. The KBr powder is used to abrade the surface and remove a layer 50–100 Å thick. The KBr is then pressed into a pellet from which good spectra are obtained. If desired, the treatment may be repeated and the various strata examined individually. An abrading machine can be used to ensure uniform removal of the surface. Potassium bromide mixed with steel wool that is later removed with a magnet will give faster cutting action. Examples given include detection of hydrocarbons on glass, phthalate ester on stainless steel, and amides on polyethylene. Reasons for adhesion failure in paint were also studied. A spray abrasion method, using powdered KBr, has been suggested for depth-profiling the interior surfaces of arteries and veins [71].

Considerable difficulty is likely to be encountered in sampling tough film-like or rubbery insoluble solids. One elegant technique is to grind the material in a mechanical grinder after freezing it with liquid nitrogen [112, 117]. The steel grinding cylinder, containing the sample and two stainless-steel balls, is placed in liquid N_2 until boiling ceases. The cylinder is then placed in a mechanical vibrator–grinder and is vibrated for a short time. In this way materials such as epoxy resin, cork, tobacco, cellulose, paper, and fingernails are reduced to a fine powder that can be incorporated subsequently into a KBr disk. Spectra of materials prepared in this manner are of excellent quality.

The KBr-pellet technique has some severe shortcomings that are not always fully recognized. Most serious, perhaps, are the problems involving changes of either the crystal structure or the composition of the sample. Solids showing polymorphism will give differences in their spectra that vary with the grinding and pressing technique (Fig. 4.4) [7]. Changes in the spectra of phenols and organic acids are apparently caused by adsorption of the molecules onto the alkali halide particles [116]. Samples may react with atmospheric H_2O or CO_2 during preparation of the pellet, although ways of circumventing this problem can be found [89, 110]. Partial or complete ion exchange, particularly for inorganic salts and for hydrochloride salts of organic amines and other bases, may cause spectral changes that completely preclude any useful purpose for the spectrum. This problem has been related to moisture absorption, time and temperature of acid–salt contact, and KBr particle size [18, 28]. Another study concluded that the free energy change of the reaction was a reliable criterion for predicting ion-exchange behavior [85].

Water absorptions invariably appear near 1640 cm^{-1} and 3450 cm^{-1}, although vacuum drying helps reduce them [7]. Stored pellets often become cloudy because of the formation of microcracks [88].

It would seem wisest to reserve the KBr mull method for samples that: (1) are insoluble in common IR solvents, (2) are amorphous or have stable crystal structures, and (3) do not contain exchangeable ions.

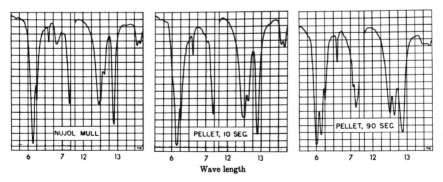

Fig. 4.4 Grinding-induced polymorphism in α-naphthalene acetamide. Left, mineral oil mull. Center and right, KI pellet after grinding in a mechanical vibrator [7]. Reprinted with permission of American Chemical Society.

ATTENUATED TOTAL REFLECTANCE

A useful and widely applicable technique for obtaining spectra of difficult samples such as rubber, foods, or cured resins is ATR. Although the phenomenon was first observed by Newton, it was not until the 1960s that possible applications to IR spectroscopy were perceived [33, 34, 59]. That decade saw rapid advances in the technique and a proliferation of commercial equipment that made the method readily accessible to spectroscopists. Achieving optimum results, however, requires some understanding on the part of the user of the physical principles involved. We discuss these principles briefly; those desiring more rigor and detail are referred to the monograph by Harrick [59].

Principles

We first consider what happens when a beam of radiation passes through an interface between two transparent materials of different refractive indices. For a beam perpendicular to the interface (beam a of Fig. 4.5), the radiation will be partly transmitted and partly reflected, with the reflectivity R (the fraction of the radiation reflected) given by

$$R = \frac{(n_2 - n_1)^2}{(n_2 + n_1)^2} \tag{4.1}$$

where n_1 and n_2 are the refractive indices of the two media at the interface.

We next examine beam b; beam b'' is reflected at the interface and the

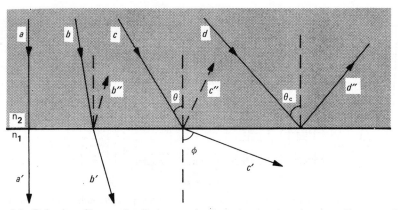

Fig. 4.5 Behavior of beam of radiation passing from dense to less dense medium at various angles of incidence to interface.

transmitted ray b' is refracted according to Snell's law:

$$n_1 \sin \theta = n_2 \sin \phi \tag{4.2}$$

Ray c behaves similarly. As the angle θ increases, at some value θ_c, the angle ϕ increases to 90°, and no radiation passes the interface. The angle θ_c is called the *critical angle*, and for all values of $\theta \geq \theta_c$, internal reflectance occurs. The critical angle can be calculated:

$$\theta_c = \sin^{-1} \frac{n_2}{n_1} \tag{4.3}$$

It might be noted that internal reflectance is virtually perfect in contrast to external reflectance (as at a first-surface mirror), where several percent of the incident radiation is lost at each reflection. Thus an internally reflected beam can undergo thousands of reflections with no loss except from absorption by the medium. The behavior of the internally reflected radiation differs slightly for its parallel and perpendicularly polarized components (Fig. 4.6).

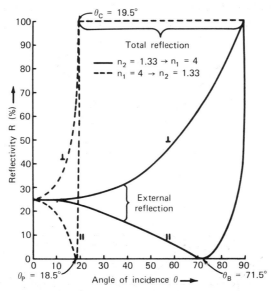

Fig. 4.6 Reflectivity versus angle of incidence for interface between media having indices $n_1 = 4$ and $n_2 = 1.33$, for light polarized perpendicular ($R\perp$) and parallel ($R_{\parallel}$) to the plane of incidence for external reflection (solid lines) and internal reflection (dashed lines); θ_c is critical angle, θ_B is Brewster's angle; and θ_p is the principal angle (θ_p is the complement of θ_B). Reproduced with permission of Interscience Publishers, from Harrick [59].

The use of ATR in spectroscopy is based on the fact that although complete internal reflection occurs at the interface, *radiation does in fact penetrate a short distance into the rarer medium* (Fig. 4.7a). This penetrating radiation, called the *evanescent wave*, can be partially absorbed by placing a sample in optical contact with the dense medium (which we call the *prism*), at which point reflectance occurs. The reflected radiation can yield an absorption spectrum that closely resembles a transmission spectrum of the sample (Fig. 4.8). This result is not automatic, however; the actual pattern obtained depends on several parameters, including the refractive indices of the prism and of the sample, the angle of incidence of the radiation, the thickness and the area of the sample, the number of reflections, and the wavelength of the radiation. How to optimize these variables to obtain the best transmission-like spectra is the subject of the paragraphs that follow.

Practice

Attenuated-total-reflectance spectroscopy is carried out by placing the sample snugly against the surface of a prism or multireflection element (Fig. 4.7b) through which the spectrometer radiation is directed by suitable transfer optics. The prism is made from a material of high refractive index such as AgCl, KRS-5, or Ge (see Table 4.1). The prism material

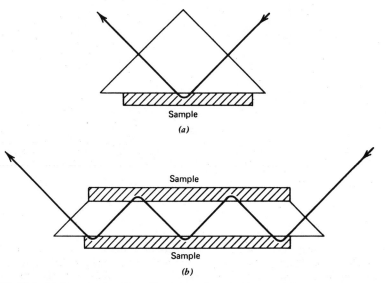

Sample

(a)

Sample

Sample

(b)

Fig. 4.7 Schematic representation of attenuated total reflectance in: (*a*) single reflection prism; (*b*) multireflection element.

TABLE 4.1 Typical Properties of Optical Materials Used for Internal Reflection Elements

Material	Useful Range[a] (in μm)	Mean Refractive Index	Critical Angle	Knoop Hardness	Reflection Loss (2 Surfaces)	Remarks
Silver chloride	0.4–20	2.0	30°	9.5	20%	Light-sensitive; reacts with metals
Silver bromide	0.5–30	2.2	27°		26%	
KRS-6, thallous bromide chloride	0.4–32	2.2	27°		26%	
Diamond	1–3.8 5.9–100	2.4	25°	7000	31%	Type II-a
KRS-5, thallous bromide iodide	0.6–40	2.4	24.6°	40.2	31%	
Zinc selenide (Irtran-4)	0.5–15	2.4	24.6°	150	31%	Brittle; releases H_2Se with acids
Cadmium teluride (Irtran-6)	1–23	2.7	22.2°	45	38%	
Arsenic selenide	1–12.5	2.8	20.9°		39%	Brittle; attacked by alkali
Silicon	1.1–6.5 33–150	3.4	15.6°	1150	52%	
Germanium	2–12	4.0	14.5°	600	59%	

[a] Thickness is 2 cm, except in diamond, which is about 1 cm.

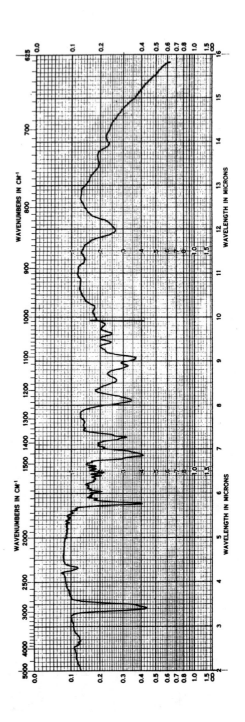

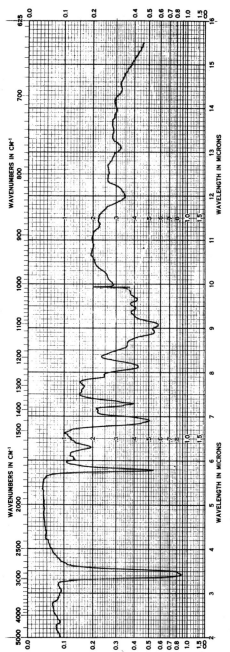

Fig. 4.8 Comparison of ATR spectrum (top) and transmission spectrum (bottom) of masking-tape adhesive.

89

must be nonabsorbing at path lengths of several centimeters and should also be tough, capable of taking a high polish, and chemically inert.

Among the most important factors affecting the spectrum are the refractive indices of the prism and sample. The refractive index of any substance undergoes a striking change at an absorption band, as shown in Fig. 4.9. If n_p is the refractive index of the prism and n_s that of the sample, and $n_{sp} = (n_s/n_p)$, $(n_{sp} < 1)$, the larger n_{sp} (i.e., the closer the refractive indices of prism and sample), the deeper is the penetration of the evanescent wave and the higher is the spectral contrast. At the same time, however, strong sample absorptions may give regions in which the refractive index of the sample becomes greater than that of the prism, and the radiation will no longer be internally reflected (Fig. 4.9). As a result, the spectrum will be distorted, unless the sample is very thin—less than the penetration depth of the evanescent ray.

Attenuated-total-reflectance prisms can be obtained in a variety of geometries and entrance angles (angles of 45° and 60° are common). In any case, the angle of incidence θ must be greater than the critical angle θ_c for total internal reflection to occur. The fact that the critical angle varies across an absorption band suggests that θ should be appreciably

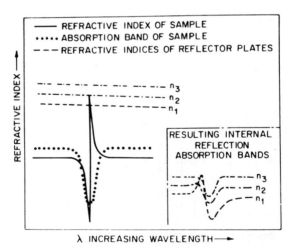

Solid Line – Refractive index of sample.
Dotted Line – Absorption band of sample.
Dashed Lines – Refractive indices of reflector plates.

Fig. 4.9 Effect on refractive index of ATR prism on absorption band (dotted line) having refractive index n_s (solid line) that approaches prism index n_i (dashed lines). The resulting internal reflectance absorption bands are shown for $n_s < n_3$; $n_s = n_2$; $n_s > n_1$. Courtesy American Society for Testing and Materials [2]. Copyright 1977.

greater than θ_c. Choosing a very large θ for general use is not appropriate, however, as penetration depth decreases with increasing θ. As a rule of thumb, Wilks and Hirschfeld have suggested [127] the addition of 0.2 to the refractive index of the sample when calculating the critical angle and then a 3° increment to the critical angle to compensate for beam convergence. The effect of incidence angle is shown in Figs. 4.10 and 4.11 and is discussed by Wilks [126]. It should be noted that spectra "distorted" by use of a low index element or low angle of incidence are useful for determining optical constants of materials [40, 64].

Other factors governing spectral contrast include the number of reflections (which will be higher for high angles of incidence) and the contact area of the sample. The number of reflections may be increased by using a thinner or longer multireflection element.

The penetration depth is on the order of one wavelength. It may be

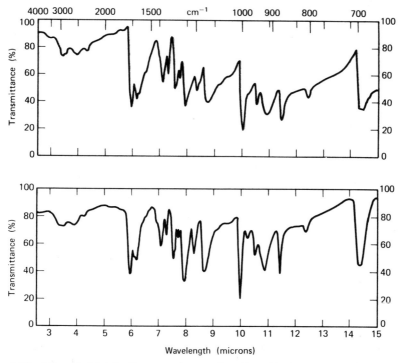

Fig. 4.10 Attenuated-total-reflectance spectra of sorbic acid with a KRS-5 element. Top, angle of incidence 45°; bottom, angle of incidence 60°. Reproduced, with permission, from *Applied Spectroscopy* [126].

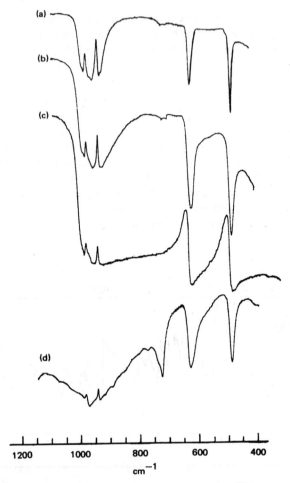

Fig. 4.11 Attenuated-total-reflectance spectra of $NaClO_3$ showing effect of angle; KRS-5 prism: (a) $\theta = 60°$; (b) $\theta = 45°$; (c) $\theta = 30°$; (d) mineral-oil mull transmission spectrum. Reproduced, with permission, from *Spectrochimica Acta* [16].

calculated using the formula:

$$d_p = \frac{\lambda_1}{2\pi(\sin^2 \theta - n_{sp}^2)^{1/2}}, \qquad (4.4)$$

where λ_1 = wavelength of the radiation in the prism = (λ/n_p). The effect on the spectrum is that bands at longer wavelengths are more intense than are those at shorter wavelengths, relative to a transmission spectrum of the same material (cf. Fig. 4.8).

Because the penetration depth can be varied by changing either the prism material, the angle of incidence, or both (and particularly if one can do scaled subtraction of the spectra using a computer), it is possible in principle to obtain a depth profile of a surface. This problem has been discussed by Tompkins [118] and by Hirschfeld [66], who claims that under favorable circumstances, individual depth elements of $\lambda/10$–$\lambda/20$ can be resolved at depths up to about 2λ within a sample.

It should not be inferred, however, that the evanescent ray terminates abruptly after some finite penetration into the rarer medium. The magnitude of its electric field drops exponentially, as shown in Fig. 4.12, and d_p is defined simply as the depth at which the field reaches $1/e$ of its initial magnitude.

In everyday practice, a large percentage of samples can be accommodated routinely by using a 45° KRS-5 multiple reflection element. Spectral contrast is adjusted by changing the sample contact area. If sample is limited, best use can be assured by confining it to the active

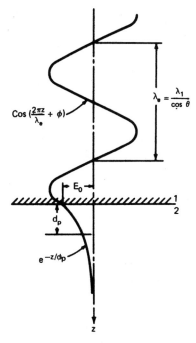

Fig. 4.12 Standing-wave amplitudes established near a surface where total internal reflectance occurs. Electric field amplitude is sinusoidal in denser medium 1 and decreases exponentially in rarer medium 2. Reproduced, with permission, from Harrick [59].

area of the element. Because the radiation seen by a dispersive spec-trmeter does not represent a homogeneous sampling of the ATR ele-ment's surface, it may be possible to find small areas of maximum sen-sitivity, or "hot spots." Active areas can be found by using narrow strips of adhering tape successively over adjacent portions of the element [92]. If possible, the sample should cover the entire width of the sensitive area. If it does not, a false zero will result, similar to the effect observed from a hole in a transmission film or a bubble in a liquid cell. Best-quality spectra are obtained by using a matching internal reflectance attachment and element in the reference beam. Such an arrangement gives reasonable compensation for atmospheric absorptions and optical losses. Because at least 20–60% of the radiation is lost at the element–air interfaces, some adjustment of the slit (or gain) may be necessary to maintain a live servo system (see pp. 40–44).

New internal reflection elements should be checked for quality by running a blank scan. The tracing should be flat at or near the theoretical value over the usable wavelength range. Poorly polished plates show scattering at shorter wavelengths. The tolerances on flatness and angles for an internal reflectance element are rather strict. For these reasons, and also because of the difficulty of obtaining highly polished flat sur-faces, it is best not to attempt repolishing of a scratched element. Rather, it should be returned to the supplier for reworking.

Sampling

Often, the biggest difficulty in using ATR is obtaining reproducible optical coupling between the reflectance element and the sample. Soft samples such as elastomers and gums or self-adhering materials usually give no problem, and band intensities are more than adequate with a multireflec-tion element. Fibers can be wound in a uniform band around both sides of the element. Flexible films, fibers, paper, and textiles can be backed with a rubber pad to ensure good contact and at the same time prevent damage to the element. Care should be taken to prevent the backing material from contacting the element and thus giving a spurious spectrum. Overtightening of the holder can be prevented by using a torque wrench or torque screwdriver.

The ATR spectra of powders are free of short-wavelength scattering and the Christiansen effects that are encountered in transmission spec-troscopy [16, 60]. Powders may be suspended in a volatile solvent and then deposited on the element by evaporation of the solvent. (Anisotropic powders can, however, orient themselves on the surface and produce anomalous spectra because of the polarization of the radiation.) Or, powders can be spread on the sticky side of adhesive tape, which is then pressed lightly against the element. Better results are obtained if the powder is reasonably finely ground and uniform in particle size, and of course the adhesive should be completely covered so as not to be visible in the spectrum.

Hard materials may present a difficult problem in sampling. If brittle, they are best ground and sampled as powders. Where a hard, flat surface is to be sampled, optical coupling can often be dramatically improved by using a thin layer of mineral oil on the surface of the sample. If mineral oil bands are objectionable, or if the surface is too rough to give a spectrum, a 0.5-mm AgCl sheet can be pressed between the element and the sample [63].

Liquid samples give good ATR spectra, and aqueous solutions have been successfully sampled this way [1]. A special element design in the form of a vertically mounted double-pass probe that can be dipped directly into liquid samples is available [57]. The depth of immersion controls spectral contrast.

Microsamples give best results when spread in an extremely thin layer over the effective sampling area of the element [43]. If the sample is less than about 1 μm thick, the spectrum will not be distorted even near the critical angle for the ATR element in air. If the sample cannot be spread in a thin layer, a beam condenser combined with a single-reflection element or the use of reflection elements having special geometries or reduced size may allow most effective use of the sample. Greatest sensitivity can be obtained by selecting conditions that give maximum penetration depth [62].

Cleaning ATR elements should be done carefully and with a minimum of contact with the polished surfaces. Solid or powdery coatings can be removed by placing pressure-sensitive tape directly onto the sample and then peeling it off. A series of solvent rinses (acetone, toluene, and methyl ethyl ketone) with the element held vertically for best drainage is usually successful. Stubborn samples may require a 30-s solvent treatment in an ultrasonic bath. The crystal should be drip dried, using a tissue to blot the last drop of solvent from the lower corner. The crystal should be handled by its edges only, and finger cots should be worn. A plasma cleaner is useful with silicon or germanium elements but cannot be used with KRS-5.

Recommended practices for ATR spectrometry are found in ASTM Method E 573-76 [2, p. 435].

GAS SAMPLING

Gas sampling usually presents no particular problem except when the sample reacts with the cell windows or body. In this case, special materials must be used for construction of the cell. Polyethylene film windows are frequently useful but cannot be subjected to very large pressure differentials. Gaskets should be avoided unless some inert material such as Viton or Teflon is used; other materials may cross-contaminate samples because of absorption and desorption [52].

Gas samples are usually pressurized to atmospheric pressure, using an inert diluent such as dry nitrogen, to minimize variations from pressure broadening (see pp. 167–169). This procedure increases sensitivity for trace constituents (except for measurements made at high resolution) and also enables quantitative gas analysis. It is wise to make some provision for stirring the gas since instantaneous diffusion may not occur. A small piece of Teflon that can be shaken inside the cell serves admirably as a stirrer.

When high sensitivity is required, as in air-pollution studies, the long path multireflection cells of White's design [124] are very useful. Cells having path lengths of up to 120 m are available commercially, and custom-built designs of over 1 km path length have been described [97]. Sensitivities of 0.1–1 ppb for pollutants such as CO, N_2O, NO, NO_2, HNO_3, O_3, and C_2H_4 have been reported [55].

Trace-component analysis in long-path cells can be badly distorted by selective adsorption or desorption of materials from the cell walls. The cell should be "rinsed" by emptying and filling it several times before the spectrum is recorded. In some cases, it may be necessary to keep fresh sample continuously flowing through the cell to obtain an accurate representation of its true composition.

Trace atmospheric contaminants such as noxious and toxic vapors may be trapped in charcoal tubes and eluted with solvent for identification by IR. A common trap design utilizes 150 mg of charcoal, through which 10 liters of air are drawn at the rate of 1 liter per minute. In one recovery study [26], 0.5 ml of CS_2 was used to extract the contaminant from the charcoal. Only about 50% recovery was obtained at room temperature, but it was found that if both the trap and the CS_2 were cooled with liquid N_2 prior to the extraction, 80–100% recoveries resulted. After 5- or 10-min extraction, the CS_2 was transferred to a 6-mm thick liquid cell (volume 0.43 ml) and a CS_2-compensated spectrum run. For the 10 compounds studied, nine could be identified at concentrations of 1 ppm or less (v/v in air).

SPECIAL SAMPLING METHODS

Reflectance

Very thin coatings (ca. 0.01 mm) on smooth metal substrates can sometimes be sampled by reflecting the radiation beam from the metal surface,
using a specular reflectance attachment. Films appreciably less in thickness than the wavelength of the radiation will give no spectrum if the
radiation is perpendicular to the metal surface because an electric standing-wave field has a node approximately at the reflecting surface. Molecules at the node cannot interact with the radiation [48]. If glancing
incidence and multiple reflections are used, however, IR spectra can be
obtained on as little as a monolayer [11, 119]. Such systems have been
used to study absorption of CO on fresh metal surfaces [119] and oxidation of metals [98].

Diffuse reflectance has not been used much in the IR region, probably
because of the experimental difficulties, but an interferometer spectrometer has become available that is specifically designed for this type of
work [128]. In the near IR, diffuse reflectance spectrophotometry has
proven useful for quantitative analysis of protein, oil, and moisture in
grains and other agricultural products [122].

Pyrolysis

If all else fails, obstinate samples often succumb to pyrolysis or dry
distillation, followed by IR analysis of the volatiles [56]. A gram or less
of the material is fragmented and placed in a borosilicate glass test tube.
The tube is held horizontally and the closed end is heated over a Bunsen
burner flame. The volatiles collect around the neck of the tube and may
be transferred easily to a salt plate for IR examination. Controlled pyrolysis using commercially available equipment is more reproducible and
also permits vapor sampling [120].

It has been found that, in most cases, spectra of the pyrolyzates
resemble those of the parent compounds. Even when they do not, the
spectra are fairly reproducible; thus reference spectra prepared in the
same manner can be used for comparison. Polyurethanes have been
identified by pyrolysis of unknown polymers [72]; fibers used in carpeting
give distinctive pyrolysis products [95].

Spectra at High and Low Temperatures

It is not generally feasible to run spectra routinely at temperatures very
different from ambient because of the time required to reach equilibrium,

but occasionally problems arise that can be resolved only by high- or low-temperature techniques. Cells now available commercially are usually more satisfactory than homemade equipment, but a number of cells designed for special problems have been described in the literature.

For temperatures only moderately above ambient, thermostated liquid cell holders, heated by a circulating fluid, or by an electrical heater, are satisfactory. If knowledge of the sample temperature is important, a fine-wire thermocouple should be inserted directly into the sample space, since readings taken at other parts of the cell may be grossly misleading. Molten solids at high temperatures may be examined in an ingenious cell that uses a platinum grid as sample support [13, 47].

Infrared spectroscopy at elevated temperatures is subject to two experimental difficulties, aside from those of achieving and maintaining the desired sample temperature. Heating of the spectrometer by a hot sample cell may occur but can be minimized by a properly insulated mounting. Second, and more serious, the sample itself acts as an IR source. This effect can be eliminated by using either a single-beam or a ratio recording double-beam spectrometer that chops the IR beam *before* it passes through the sample (the spectrometer does not respond to unchopped radiation from the sample).

The design of low-temperature cells is somewhat more difficult, since the sample must be well insulated from the atmosphere to prevent moisture condensation. This insulation usually takes the form of a vacuum. Another difficulty is thereby introduced—that of devising a vacuum-tight sample container. This problem is often avoided by precooling the cell and then spraying the sample onto a cold window, where it immediately freezes and does not contribute any appreciable vapor pressure.

A technique called matrix isolation [9, 27, 54, 84] has been used extensively to study unstable molecular species such as free radicals. Here the sample (e.g., from a small furnace) is continuously deposited on a cold window along with a much larger amount of an inert matrix gas such as argon or nitrogen. The sample molecules are frozen into the inert matrix in low concentration so they cannot interact with each other. After enough material is deposited, the spectrum is scanned.

A related technique for qualitative analysis of complex gas mixtures has been called *pseudomatrix isolation* [102, 103]. Here the sample is deposited in bursts over a short time period. The method is said to be quite sensitive for minor constituents.

A 2-m path multipass cell for spectroscopy of samples dissolved in liquid argon has been described [68].

Low-temperature spectroscopy has found some application to certain kinds of materials that give less than definitive room-temperature spectra (Fig. 4.13), such as amino acids and polypeptides [36], cellulosic materials [82], polyvinylacetate–acrylate copolymers [53]; and carbohydrates [74, 75]. It can be used to distinguish materials that have nearly identical room-temperature spectra (e.g., *n*-hexyl bromide and *n*-heptyl bromide) [17].

Temperature calibration may be carried out using liquids of known melting point [38]. A comprehensive literature review on subambient spectroscopy to 1968 has been published by Hermann [61].

A cold sample in the spectrometer beam is seen by the spectrometer as a heat sink, and unless the beam is modulated before it traverses the sample, a distorted spectrum (in the sense that the actual zero is considerably displaced) will result. The effect will be of little importance for qualitative spectra, but it should be taken into account for more accurate work.

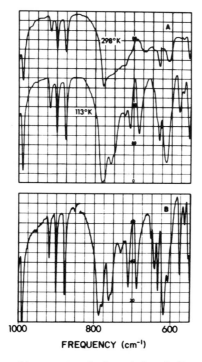

FREQUENCY (cm⁻¹)

Fig. 4.13 Partial spectra of lactose monohydrate (mineral-oil mull) at 298 K and 113 K (upper) and 20 K (lower). Less sample is needed at lower temperatures. Reproduced, with permission, from *Applied Spectroscopy Reviews* [75].

Coupled IR–Chromatography

Cuts from a gas chromatograph (GC) may be trapped and run as liquid samples [10] or examined as they elute from the chromatograph by either stopped-flow or "on-the-fly" techniques [6]. Trapping usually requires some skillful manipulation of small quantities of sample (10–100 μg) and is thus less appealing than *in situ* methods. It has the advantage, however, of producing spectra directly comparable to spectra in the reference library. Low-volatility samples can be trapped on powdered KBr in a capillary and subsequently pressed into a pellet [93].

Chromatographed fractions are usually scanned in the vapor at 180–200°C. Stopped-flow or vapor-trapping methods can be used with conventional IR spectrometers [15, 23, 81]. Sensitivity is ordinarily inadequate for minor constituents but may be improved by a special chromatograph design optimized for IR sampling [107]. Larger sample injections ($\leq$100 μl) are used, and good separation of components is achieved by operating at a pressure of 120 N/cm^2 (i.e., 180 psig) across the entire system. Pressure drop in the column is relatively small. Under the conditions used, stopping the gas flow during the IR scan does not degrade chromatographic resolution.

"On-the-fly" measurements require use of either a special rapid-scan dispersive spectrometer [73] or an interferometer spectrometer. Best sensitivity is attained when the light-pipe (cell) dimensions, flow rates, and scan times are optimized [50, 121].

Examples of GC–IR analyses using a dispersive spectrometer have been reported [41]. The spectra shown in Figure 4.14 were obtained from the heart cut of a petroleum stream. It had been determined by GC–mass spectrometry (MS) that the molecular formula of the fraction was $C_{10}H_{14}$, but the material could be characterized only as indane or an isomer of methyl styrene. Even though spectral quality is not comparable to that obtained with more leisurely scans on larger samples, the material could clearly be characterized as *m*-methylstyrene. Other examples given in the same paper highlight the need for complementary techniques such as NMR and Raman spectroscopy on trapped microsamples for solving problems of identification. Since these techniques require 0.1–1 μl, they are most valuable when adequate sample is available. They then permit quick fractionation and characterization of the components without resorting to fractional distillation.

The IR interferometer, because of its energy advantage, usually gives superior results, especially if sample size is limited. It was found in one study that only 10 ng of isobutyl methacrylate scanned "on the fly" at 8-cm^{-1} resolution with 16 signal-averaged scans gave an interpretable

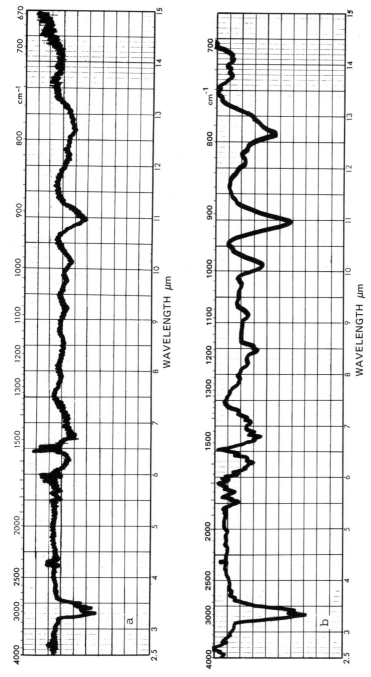

Fig. 4.14 Gas-chromatography-infrared spectra of *meta*-methylstyrene: (*a*) single 30-s scan; (*b*) sixteen 6-s signal-averaged scans. Reproduced, with permission, from *Applied Spectroscopy* [41].

101

spectrum after suitable smoothing (Fig. 4.15). The interferometer was equipped with a cooled mercury cadmium telluride detector.

Trapping the sample fraction in the cell increases sensitivity by giving a better signal : noise ratio and permits signal averaging of a larger number of scans (provided that the sample does not decompose at the temperature of the light pipe). Methods for coupling GC and IR have been reviewed by Freeman [39] and Littlewood [80], but advances occur rapidly; the recent literature is the best source of up-to-date information.

Any spectrum recorded in the vapor phase at elevated temperatures will have quite a different appearance from condensed-state room-temperature spectra (light molecules and those normally showing hydrogen bonding will show the largest changes). Band frequencies will also be shifted. Reference spectra run under similar conditions may be needed for identification [123].

Other types of chromatography have also been coupled with IR spectroscopy. Infrared identification of compounds separated by thin-layer chromatography (TLC) is difficult because of the small quantity of sample. Removal of a band of the TLC absorbent followed by extraction has been suggested as a technique for identifying the absorbed species, but

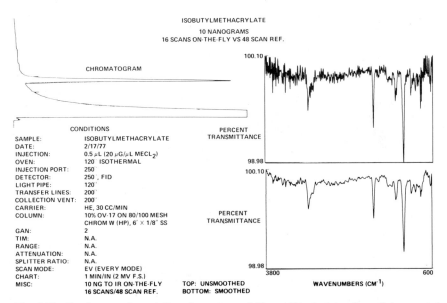

Fig. 4.15 Fourier transform–infrared spectrum of 10 ng of *iso*-butylmethacrylate scanned "on the fly." Top spectrum is unsmoothed; bottom spectrum smoothed. Reproduced, with permission, from *Applied Spectroscopy* [121].

high blanks render this procedure uncertain [25]. An *in situ* approach by Griffiths and colleagues [44, 94] using Fourier transform spectroscopy has been more successful, and detection of 1–10 μg of adsorbed pesticides was reported. The silica gel or alumina adsorbent was deposited in a 100-μm layer on an AgCl substrate. After chromatography of the sample, the spots were visualized by exposing the plate to iodine vapor. The plate was then inserted into the spectrometer and the spectra measured directly through the plate at the spot location, using a 4× beam condenser. Four hundred scans were signal averaged at 4 cm^{-1} resolution. Spectra were ratioed against single-beam spectra taken at an adjacent sample-free portion of the plate. In this way, adsorbent bands could be eliminated from the spectrum. Scattering at shorter wavelengths was markedly reduced by treating the plates with Fluorolube or mineral oil. Programmed multiple development TLC combined with a spectrometer having a more sensitive detector permitted usable spectra to be measured on as little as 10 ng of methylene blue [45].

Polymers and other materials separated by gel-permeation chromatography (GPC) or similar procedures can be examined by evaporating the solvent from selected cuts and treating the residue as a microsample. Infrared detection of GPC fractions can be accomplished by using a 1-mm-thick microcell and setting the spectrometer at a fixed wavelength [24]. The sample must be reinjected if detection at another wavelength is required.

Microsampling

Handling technique becomes particularly important when microgram quantities of sample are to be analyzed [12]. There is no substitute for practice to attain proficiency, and time spent making trial runs on known samples before manipulating the unknown material will not be wasted. Recommended practices for IR microanalysis are covered in ASTM Method E-334-68 [2].

Microsampling accessories of various types, such as beam condensers, small-volume gas cells, liquid microcells, and microdies for KBr pellets, are available from the instrument manufacturers. Satisfactory equipment can often be improvised in the laboratory.

When working with microsamples, the analyst must consider how to optimize energy transmission so as to obtain the most useful spectrum. Should he concentrate the sample in a small area and reduce the beam diameter, or should he spread the sample over a larger area to avoid vignetting the beam and use ordinate expansion to increase the spectrum intensity? The answer depends to some extent on the type of spectrometer he is using and whether the output permits digital manipulation of absorbance data. Generally speaking, however, the beam area intercepted by the sample should be as small as possible, consistent with the requirement that the beam fill the slits at their maximum width. Some energy loss can be tolerated from vignetting at the top and bottom of the slits; this loss will be essentially wavelength independent and can be compensated for by using a reference beam attentuator and widening the slits to restore lost servo energy. Failure to fill the full *width* of the slits, however, will result in a wavelength-dependent energy loss that will cause sluggish pen response in some regions and possible excursions of the background. A beam condenser, which reduces the beam size at the sample position without significant energy loss, can be a useful aid in examining small samples.

Ordinate (pen) expansion can be used to amplify the spectral pattern, but because noise is increased by the same factor as the bands, the slits must be opened by the square root of the expansion factor and the gain decreased correspondingly. Background slope and other artifacts are also increased, so extended scans are rarely possible with ordinate expansion. If the spectrum is digitized, however, accurate background corrections can be made and ordinate expansion becomes a practical option. The process is described more fully later in this section.

The KBr pellet technique is used extensively for microsampling of nonvolatile solids [5, 77]. Lyophilization is usually employed to disperse the sample in the KBr, since grinding and quantitatively transferring a few micrograms of material is difficult, to say the least. Furthermore, grinding and other manipulations introduce impurities at a level far greater than the sample concentration. It should be reemphasized that manipulation of microgram quantities of sample, whatever the sampling technique, must be done only in scrupulously clean equipment and with utmost care.

A system for concentrating microsamples in a KBr matrix has been devised [42]. A triangular piece of KBr made from lightly sintered powder (Wick-Stick®) is placed in a glass vial, point up, and the sample and solvent are added. Both migrate into the KBr, and final evaporation occurs on the top of the KBr element. The sample remains at the tip, which is then broken off and formed into a micropellet. Spectra can be obtained from about 50 μg of material, and the sample is subjected to only minimal handling.

Microsampling of even slightly volatile materials using the KBr pellet method is often unsatisfactory because of sample loss by evaporation. Griffiths and Block [51] tested a series of phosphonate acid esters having a vapor pressure of about 10^{-4} torr at 20°C. They demonstrated that a large fraction of the sample was lost by evaporation during transfer of the KBr and sample to the microdie, and over 90% of the material disappeared in a few minutes' time. In a similar experiment, King [76] studied loss of 2,6-dimethoxyphenol, which has even lower volatility than the phosphonate acid esters. He found that when a solvent (CS_2) is used as sample carrier, with the solvent evaporated on the KBr powder, the amount of loss is a function of the sample dilution. The more dilute solutions showed the largest losses. In any case, more than 10 ng of sample was required to obtain a spectrum even using an interferometer spectrometer with ordinate expansion. In practice, it may be concluded that for even slightly volatile materials, the sensitivity potential for dilute solutions in a small-volume cell is greater than for micro KBr pellets.

A liquid microcell has been described that permits solution spectra of a 10–100-μg sample without a beam condenser [114]. Solid samples, such as fibers or nonvolatile liquids, may also be examined without any preparation other than mounting, if a beam condenser is used. The use of watch jewels for sample holders has been proposed [4].

The energy advantage of the interferometer spectrometer permits the use of smaller samples. An aperture of a few tenths millimeter diameter punched in brass shim stock can be used to mount tiny samples for examination. The optimum diameter of the microsampling cell for Fourier transform IR has been derived by Hirschfeld [65].

Infrared microsampling below the l-ng level has been accomplished by using a Fourier transform spectrometer with an 8× beam condenser [22]. Sample preparation is done under a microscope, as samples are frequently not visible to the unaided eye. The sample is supported on a thin NaCl plate and is centered within an aperture, 50–200 μm in diameter, punched in thin brass shim stock. The size of the aperture is chosen such that it is completely filled by the sample. Optimum sample thickness is a few micrometers, and various methods are used to reduce the thickness if it is necessary to do so. The need for extreme care and cleanliness is stressed, as a single dust particle is often 10–100 times the weight of the samples being studied. Figure 4.16 shows the spectrum of 6 ng of triphenylphosphate, scanned 1000 times, and Fig. 4.17 shows 0.9 ng of cellulose acetate with 90 pg of triphenylphosphate, scanned 96,000 times (40 h of data accumulation). It is expected that instrumental and technique improvements will permit study of even smaller samples. The fact that the IR spectra of crystalline particles (both band positions and intensities) vary with particle size should not be forgotten in work of this type, however (see pp. 251–252 of present work; also Ruppin and Englman [104].

Enhancement of computerized data from ratio recording or interfero-

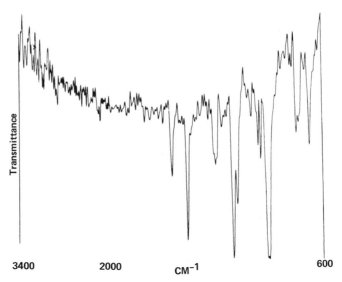

Fig. 4.16 Spectrum of 6 ng of triphenyl phosphate in a 100-μm aperture, scanned 1000 times at 8-cm^{-1} resolution [22]. Reproduced with permission of American Chemical Society.

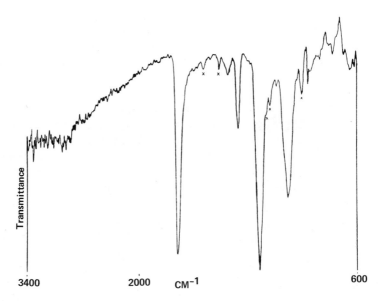

Fig. 4.17 Spectrum of 0.9 ng of cellulose acetate containing 10% triphenyl phosphate, scanned 96,000 times at 8-cm^{-1} resolution [22]. Reproduced with permission of American Chemical Society.

106

meter spectrometers can add another dimension to the practice of IR microanalysis. Conventional optical null spectrometers, even with digitized outputs, may be less satisfactory because of the mechanical problems associated with the optical wedge (inertia, backlash, and lack of energy near zero transmission). Digitized data is manipulated in a computer by use of four basic routines: absorbance expansion, digital smoothing, absorbance subtraction, and spectral accumulation [19].

The process basically is as follows. A sample is run and digitized using standard or semimicrotechniques, and a "blank" run is also made. Even though recordings of the two runs may appear to be identical, the sample record contains spectral information about the sample. The blank is subtracted from the sample spectrum to correct for solvents, sloping background, impurities, and so on. This record is expanded by an appropriate factor—perhaps as much as 100 or 200. Since the noise level is also expanded, a digital smoothing routine must be applied to recover a spectrum of quality nearly equal to that of a normal spectrum. Further reductions in noise level may be obtained by successive accumulation and signal averaging of sample spectra. Regions of strong solvent absorption are, of course, blacked out and cannot be recovered. Some examples of microanalysis done using a ratio-recording dispersive spectrometer interfaced to a laboratory minicomputer are shown in Figs. 4.18 and 4.19.

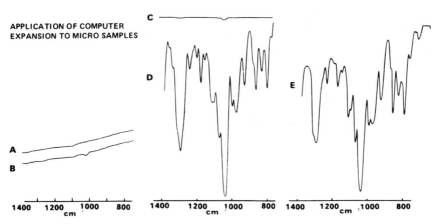

Fig. 4.18 Microanalysis using spectrometer interfaced to minicomputer: A, background trace from ATR element; B, 1 μg of Birlane deposited on ATR element; C, computer-calculated difference $B - A$; D, spectrum C expanded 200× and smoothed; E, reference ATR spectrum of Birlane. Reproduced, with permission, from *American Laboratory* [19]. Copyright International Scientific Communications.

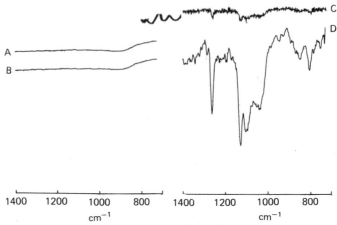

Fig. 4.19 Illustration of spectrum accumulation by multiple scans: *A*, reflectance spectrum of metal surface; *B*, same surface treated to have an estimated 20-Å-thick layer of silicone oil; *C*, result of 15 accumulations of *B*; *D*, spectrum *C* expanded 10× and smoothed. Reproduced, with permission, from *American Laboratory* [19]. Copyright International Scientific Communications.

Spectra at High Pressures

The behavior of materials under pressures up to 10,000 atm has been studied spectroscopically by a number of workers. Such studies present many special problems, particularly in cell design. The best window material found to date for extreme pressure cells is diamond, of which the so-called Type II-a diamonds are quite transparent in the IR region. The optical aperture is understandably small, so beam condensers must be used. Another problem is that of calibration of the internal cell pressure. These problems and others are discussed by Ferraro and Basile [37], who review cell design, spectrometers, windows, and calibration and include an extensive bibliography.

Emission Spectra

Obtaining a spectrum from a sample held only a few degrees above room temperature using its own emission may seem farfetched, but this experiment has been done successfully using interferometer spectrometers. Liquids and solids are studied by depositing a thin film on a specular mirror substrate having good reflectivity. The basic difficulty is that the sample must be not only very thin but in the right range of thickness, or the outer layers will reabsorb the energy emitted from the inner sections.

Thermal gradients also cause problems. For a sample of the proper thickness, the bands that normally absorb will be seen in emission; in other words, the spectrum will be inverted. Thick samples give profiles that approximate a blackbody, with strong bands no longer in evidence. Gases produce good emission spectra under proper conditions [49]. Emission spectroscopy has also been done with conventional dispersive spectrometers, but with greater difficulty and with poorer results [32].

A clever optical arrangement used by Lauer and Peterkin [78, 79] allowed them to study the behavior of a thin lubricating film under load using its emission spectrum, which was directed into an interferometer spectrometer.

Absorption spectroscopy almost invariably gives better results than the emission technique, except possibly for very weak absorbers [67] and should hence be the preferred approach when possible.

Aqueous Solutions

Water, because of its highly polar nature, is a good solvent for polar molecules. However, it is also a poor IR solvent because its absorptions are very intense, and its application thus has been restricted to special situations such as biological studies [115]. A large portion of the "fingerprint" region is available, however, if cell thicknesses on the order of 10–25 μm are used. The intensities of the bands near 3200 cm^{-1} and below 800 cm^{-1} are such as to preclude the use of these regions, except at very short paths, 2.5 μm or less, which are difficult to achieve and almost impossible to reproduce. Thus the useable range is 800–3100 cm^{-1}, with perhaps some difficulty around 1640 cm^{-1} (Fig. 4.20). (The range may be expanded somewhat by use of deuterium oxide). Also, at such

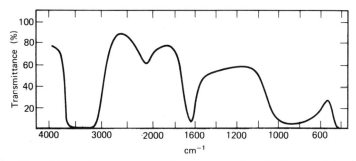

Fig. 4.20 Normal transmission spectrum of water, approximately 12 μm path length, AgBr windows [20]. Reproduced with permission of Heyden & Son, Ltd.

short path lengths, sample absorptions are very weak, particularly in dilute solutions.

Computer enhancement of such spectra may, however, make infrared analysis of aqueous solutions practical, if not routine [20]. The techniques are the same as previously described for use with microsamples. First, background absorption from pure water is subtracted from the sample spectrum. Care must be taken to achieve temperature equilibrium, as the water spectrum is quite temperature sensitive (Fig. 4.21). Also, it must be remembered that water interacts strongly with other polar molecules to influence the spectra of both, so solvent–solute interaction may result in imperfect compensation. Next, the digitized spectrum is converted to absorbance, multiplied by an appropriate expansion factor, converted back to transmittance, and finally smoothed. Figure 4.22 shows the spectrum of an aqueous solution of a pharmaceutical product. Bands due to sodium citrate, sodium acetyl salicylate, and dissolved CO_2 (2350 cm^{-1}) are apparent. These spectra were recorded using a dispersive (ratio recording) spectrometer with a digitized output and attached data processor.

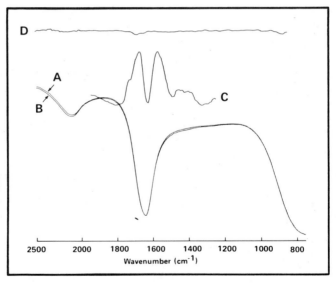

Fig. 4.21 Effect of sample temperature on computer-subtracted water spectra: A, cell at ambient temperature; B, cell equilibrated in spectrometer; C, difference $B - A$ (10× expansion); D, difference between two runs equilibrated 20 min each [20]. Reproduced with permission of Heyden & Son, Ltd.

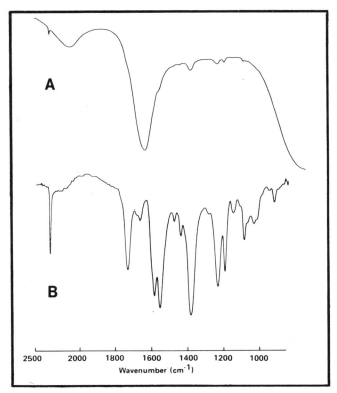

Fig. 4.22 Aqueous solution spectra: *A*, solution of soluble aspirin; *B*, spectrum *A* after subtraction of water background and 15× expansion [20]. Reproduced with permission of Heyden & Son, Ltd.

CELL CONSTRUCTION AND MAINTENANCE

Fixed-thickness Liquid Cells

The usual IR cell assembly consists of a metal frame containing ports for filling and emptying the cell, two IR-transmitting windows, one of which is drilled to admit solution to the sampling space, and a shim or gasket, which separates the windows and determines the thickness of the sample space. Gaskets of lead or a solvent-resistant polymer may be used to ensure a leak-proof seal between the cell frame filling holes and the drilled window. Frames are usually plates of brass, stainless steel, or aluminum. Gaskets are ordinarily made from freshly amalgamated lead, which shortly becomes quite hard and adheres well to the window surface. Other amalgamated metals (e.g., thin copper or brass shim stock) may be used, as well as gold, silver, indium, or Teflon sheet.

The many types of liquid sample cells commercially available differ widely in several important respects. Factors to consider in selecting a particular design for use in the laboratory include the following: (1) whether it can be filled easily, without bathing the exterior surfaces of the windows with solution, (2) whether it can be conveniently emptied and flushed, (3) whether it is possible to disassemble the cell, clean or repolish the windows, and reassemble the cell in the laboratory, (4) whether the cell volume is compatible with the amount of sample ordinarily available, and (5) whether the cost of the cell is reasonable in light of the use to which it will be put. It is suggested that several models be tried before one is adopted as standard.

A variety of materials, differing in characteristics such as transmission range, hardness, workability, cost, and refractive index, is available for windows. Transmission is important for obvious reasons. A thin window will transmit longer-wavelength radiation than a thick one, enabling the use of cells over a slightly wider wavelength span than a prism of the same material. A thin KBr pellet, for example, transmits satisfactorily to 250 cm^{-1}, whereas the cutoff point for a KBr prism is about 400 cm^{-1}.

Hardness and workability influence the ease with which the window may be repolished if the surface is fogged or marred. Along the alkali halide crystals, NaCl is quite easily polished to give a flat, transparent window; CsBr, on the other hand, is softer, less brittle, and much more difficult to polish.

Refractive index is important because reflection losses from the surfaces vary with refractive index, according to the expression:

$$R = \frac{(n_2 - n_1)^2}{(n_2 + n_1)^2} \tag{4.5}$$

where n_1 and n_2 are the refractive indices of the two media at an interface. For NaCl, with n approximately 1.53, $R = 0.04$, or about 4% per surface in air. On the other hand, AgCl has n of about 2 and $R = 11\%$ per surface. High-index optical materials are sometimes coated to reduce reflection losses (e.g., Ag_2S on AgCl), but the coating is usually fragile. A germanium cell coated with an antireflection coating has been described for use with aqueous solutions [29]. Cells with high-index windows show interference fringes superimposed on the spectrum of the sample even though the cell is filled with liquid (see pp. 236–238).

Demountable cells, which consist of a holder, two windows, and a gasket, are occasionally used for viscous liquids or other substances not amenable to sampling in a conventional liquid cell. Reproducibility of thickness is naturally not very good but may be adequate for qualitative purposes.

Samples that attack salt cell windows may be run in cells with inert window materials. If such cells are not available, such samples may be presaturated with powdered salt or sealed into polyethylene bags, which are then mounted between salt plates.

A summary of the properties of some commonly used window materials is given in Table 4.2. More detailed and precise data are given by Smith [108]. Transmission characteristics of materials useful for spectroscopy in the 30–500-cm^{-1} region are given by Fateley, Witkowski, et al. [35].

Liquid cells should be designed so that they can be flushed with clean solvent and dried easily. Viscous or gummy samples may require several successive cleanings with solvent flowing in alternate directions. Residual solvent is easily removed by blowing *dried* compressed air through the cell. If the air is not well dried, the cooling effect of the evaporating solvent may condense moisture within the cell and fog the windows.

Variable-thickness Liquid Cells

Variable-path liquid cells are useful principally as compensator cells in difference spectroscopy. The thickness of a sample cell is seldom critical, but for precise solvent compensation, the thickness of the reference cell may have to be adjusted carefully to conform to both the thickness and the concentration of the sample layer.

Several types of variable-path cells have been described in the literature, and some are available commercially. The most complicated type is that in which one window is fixed and the other is moved by an annular screw-thread arrangement [21, 125]. The demands on flatness and planarity of the windows and on mechanical precision are very great in this design.

A much simpler cell consists of two windows separated by a wedge-shaped spacer. By varying the lateral position of the cell in the beam, one can empirically select an appropriate degree of compensation. Another approach is to hot-press a wedge-shaped cavity into a rock-salt block [8].

Gas Cells

Cells for gas sampling can be made very simply from large-diameter glass tubing with an attached stopcock for evacuation. Windows may be sealed on with beeswax–rosin mixture, glyptal, epoxy, or other cement. One design uses a pressure seal in which the window is held against an O-ring

TABLE 4.2 Properties of Some Commonly Used Cell Window Materials

Window Material	Chemical Composition	Transmission Range (in μm)	Sensitive to	Refractive Index	Remarks
Glass	—	0.35-2	HF, alkali	1.5-1.9	
Quartz	SiO_2	0.2-4	HF	1.43	
LiF	LiF	0.2-7	Acid	1.39	
Sapphire	Al_2O_3	0.2-5.5	—	1.77	Good strength; no cleavage
Fluorite	CaF_2	0.2-10	NH_4^+ salts	1.40	Insoluble in water
Irtran I[a]	MgF_2	2-8	—	1.3	Polycrystalline
Servofrax[b]	As_2S_3	1-12	Alkali	2.59	Softens at 195°C.
BaF_2	BaF_2	0.2-13		1.45	Insoluble in water
Rock salt	NaCl	0.2-16	Water, glycerine	1.52	
Irtran II[a]	Zn_2S	1-14		2.24	Insoluble in most solvents
Sylvite	KCl	0.3-21	Water, glycerine	1.49	
Irtran III[a]	CaF_2	0.2-11		1.34	Polycrystalline; no cleavage
Irtran IV[a]	Zn_2Se	1-21		2.5	Polycrystalline
KBr	KBr	0.2-27	Water, alcohol	1.53	Hygroscopic
AgCl	AgCl	0.6-25	Metals, light	2.00	Very soft
Ge	Ge	2-20	—	4.0	
Si	Si	1.5-?	HF, alkali	3.4	Long wavelength limit depends on purity
KRS-5	Tl_2BrI	0.7-38	Alcohol, HNO_3	2.38	Toxic, soft
CsBr	CsBr	0.3-40	Water, alcohol	1.66	Hygroscopic, soft
CsI	CsI	0.3-50	Water, alcohol	1.74	Hygroscopic, soft
Polyethylene	$(CH_2CH_2)_n$	20-200		1.52	Very soft

[a] Trademark of Eastman Kodak Company.
[b] Trademark of Servo Corporation of America.

gasket by a threaded annular ring so that the windows may be removed easily for cleaning and polishing.

Cell Maintenance: Repolishing Optical Materials

It is inevitable that cell windows will sooner or later accumulate deposits or become fogged, scratched, or etched. Deposits are detected by periodically running solvent-compensated cell matches because visible deposits may not cause IR absorption, and IR-absorbing accumulations may not be visible to the eye. Such precipitates can occasionally be removed by allowing a solvent such as dioxane, acetone, or dimethyl formamide to stand in the cell for a few hours. More often, however, the cell will have to be disassembled and the windows repolished.

Cell disassembly is usually not difficult, although some rather persistent working with a sharp razor blade may be necessary to separate the windows from the amalgamated spacer (no attempt should be made to save the spacer as it cannot be reused). It is wise to clean the filling ports and the remainder of the cell frame before reassembling the cell.

Whereas some practice is needed to attain proficiency in the art of repolishing windows, the technique is not difficult and should be mastered by every practicing spectroscopist. (The polishing of prisms is more demanding, however, and is best left to the professional optician). If initial grinding is necessary, it can be accomplished on a flat glass disk using abrasive powder, with water or kerosene as a lubricant. Usually only one fine grade of abrasive (500 mesh) is required. Alternatively, one may use dry abrasive paper mounted on a flat glass surface, taking care not to rock the crystal so that the ground surface becomes convex. After grinding is complete, the window should be carefully cleaned of all traces of grit, because any small particles of abrasive remaining will make their presence known by producing scratches during the polishing step.

There are several methods for repolishing alkali halide crystals. Since these materials are soft (compared to glass), they can be polished comparatively rapidly to produce surfaces that are inferior to glass optics but perfectly satisfactory for IR use. The easiest method to learn, but the one that is least likely to give a flat window, is the cloth lap technique. The cloth lap is made by tightly stretching a piece of hard, dense cloth, such as fine silk, over a glass disk or other flat surface. The cloth is moistened slightly with a thin slurry of an optical polishing agent such as rouge, Barnsite, Aloxite, or Shamva with water, ethylene glycol, or ethanol. The crystal is rubbed on the lap until the window surface is smooth and is then moved quickly to a dry spot on the lap, where it is polished by further rubbing. Very clear surfaces can be generated by this method, but they inevitably will be slightly rounded.

Another method utilizes a glass flat with a ground surface as a lap. The glass is moistened slightly by breathing on it or with an atomizer, and the salt is immediately polished on the surface. This method can give quite flat plates, but clarity may be slightly inferior to that obtained by other methods.

A third method, employing a pitch lap, is more difficult to master than the preceding two but is capable of giving flat plates of good optical quality. The lap is prepared from optical pitch and beeswax [99]. In this method the surface of the pitch pad is barely dampened with a thin slurry of water and optical polishing compound, and the crystal is rubbed against its surface with a circular motion. After a short time the crystal is removed from the lap and immediately is polished on a soft dry cloth. This process is continued, with frequent checks of surface flatness against an optical flat illuminated with a sodium vapor lamp, until the desired degree of flatness and clarity is obtained. By working the edges or the center of the pitch lap, one can increase the concavity or convexity of the crystal surface. The fringe pattern observed under monochromatic light as described above is actually a "contour map" that indicates the degree of success in this manipulation. A minimum number of circular fringes (say, 3 or 4) is the most desirable pattern. A large number of irregularly shaped fringes indicates an irregular surface, often the result of a pitch pad that is too wet. A large number of circular fringes indicates a curved (concave or convex) surface. The secret to obtaining flat plates is to use a flat pitch lap. The lap is flattened by wetting it and then putting a weighted glass flat on its surface for 1–2 h. This operation may need to be repeated at intervals during the polishing process, so if many plates are to be polished it is convenient to have two pitch pads, which can be used alternately.

To prevent fogging of the polished surfaces by moist fingers, thin rubber finger cots (sold by medical-supply houses), which are more comfortable than rubber gloves, should be worn. It bears repeating that the polishing area should be carefully protected from carryover of grit from the coarse grinding operation.

Proficiency in operations such as cleaving, drilling, and cutting alkali halide crystals is acquired by experience.

Harder crystals (e.g., CaF_2, BaF_2, and LiF) can be polished by the technique outlined in the preceding paragraphs for alkali halide crystals, except that concentrated HCl is substituted for water [31]. In this case the polishing operation should be carried on in a hood, with due respect for the properties of concentrated HCl. Silver chloride reputedly can be polished with the aid of sodium thiosulfate solution or butyl amine in ethanol [86]. When using silver chloride, one should remember that it reacts readily on contact with metals less noble than silver, to the considerable detriment of both the silver chloride and the metal.

Reassembly of the cell is carried out in logical sequence with proper regard for polished window surfaces. The important points to remember during this step are: (1) the cell spacer must be carefully flattened before assembly in order to eliminate metal burrs or chips; (2) the spacer, if made of lead, should be amalgamated with lead-saturated mercury; and (3) care should be exercised in assembly to make sure filling holes are properly aligned.

After the amalgam has hardened (3-4 h), the cell should be checked for leaks by filling it with a volatile liquid. Leaks are usually due to insufficiently flat windows and/or spacer, and cannot be satisfactorily repaired. Leaky cells should be rebuilt.

Polished alkali halide surfaces are extremely perishable, and cells should be stored in a dry atmosphere such as a desiccator, or in a storage cabinet kept a few degrees above room temperature by a small electric light bulb.

Determination of Cell Thickness

Thickness of cells in the range of 0.03–0.6 mm is determined most accurately by the method of interference fringes [109]. This method takes advantage of the maxima and minima produced in the spectrum of an empty cell, resulting from reinforcement and destructive interference of radiation reflected from the internal surfaces of the cell windows (Fig. 4.23). The distance between windows can be calculated precisely from the equation

$$t \ (\mu\text{m}) = \frac{n}{2} \cdot \frac{\lambda_1 \lambda_2}{\lambda_1 - \lambda_2} \ \text{or} \ t \ (\text{mm}) = \frac{n}{2} \cdot \frac{10}{\nu_1 - \nu_2} \tag{4.6}$$

where t is the cell thickness, λ_1 and λ_2 are the wavelengths of two maxima, ν_1 and ν_2 are the corresponding frequencies (cm^{-1}), and n is the number of maxima between λ_1 and λ_2. Cells thicker than 0.6 mm may be characterized by measuring the gasket thickness with a micrometer.

A less satisfactory method, but one that may be used if interference fringes cannot be obtained, is to use some standard liquid (e.g., CCl_4 or C_6H_{12}) in the unknown cell, and compare the absorbance of a suitable band with that obtained for the same liquid in a cell of known thickness [96]. The band used for comparison should be broad and have about 30–

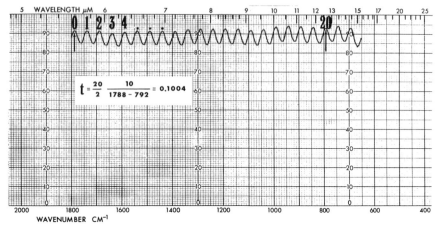

Fig. 4.23 Calculation of cell thickness from fringe pattern. Note that first fringe is counted as zero.

50% transmittance, and both cells should be run at the same time to minimize longer-term spectrometer errors.

References

1. Ahlijah, G. E. B. Y., and E. F. Mooney, *Spectrochim. Acta,* **25A,** 619 (1969).
2. American Society for Testing and Materials, *1977 Annual Book of ASTM Standards,* Part 42, ASTM, Philadelphia, 1977.
3. American Society for Testing and Materials, Committee E-13, *Manual on Recommended Practices in Spectrophotometry,* ASTM, Philadelphia, 1966.
4. Anderson, D. H., and T. E. Wilson, *Anal. Chem.,* **47,** 2482 (1975).
5. Anderson, D. H., and N. B. Woodall, *Anal. Chem.,* **25,** 1906 (1953).
6. Azarraga, L. V., and A. C. McCall, *Environ. Prot. Technol. Ser.,* EPA-660/2-73-034/273-034, 1974.
7. Baker, A. W., *J. Phys. Chem.,* **61,** 450 (1957).
8. Banas, E. M., and R. R. Hopkins, *Appl. Spectrosc.,* **15,** 153 (1961).
9. Barnes, A. J., *Rev. Anal. Chem.,* **1,** 193 (1972).
10. Blake, B. H., D. S. Erley, and F. L. Beman, *Appl. Spectrosc.,* **18,** 114 (1964).
11. Blanke, J. F., S. E. Vincent, and J. Overend, *Spectrochim. Acta,* **32A,** 163 (1976).
12. Blinn, R. C., *Adv. Chem. Ser.,* **104,** 81 (1971).

13. Boland, F. G., and J. W. Milne, *Spectrochim. Acta,* **29A,** 1214 (1973).

14. Bradley, K. B., and W. J. Potts, Jr., *Appl. Spectrosc.,* **12,** 77 (1958).

15. Brady, R. F., Jr., *Anal. Chem.,* **47,** 1425 (1975).

16. Brooker, M. H., and D. E. Irish, *Spectrochim. Acta,* **28A,** 701 (1972).

17. Caspary, R., *Appl. Spectrosc.,* **22,** 694 (1968).

18. Cleverley, B., *Appl. Spectrosc.,* **30,** 465 (1976).

19. Coates, J. P., *Am. Lab.,* **8**(11), 67 (1976).

20. Coates, J. P., *Eur. Spectrosc. News* (16), 25 (1978).

21. Coates, V. J., *Rev. Sci. Instrum.,* **22,** 853 (1951).

22. Cournoyer, R., J. C. Shearer, and D. H. Anderson, *Anal. Chem.,* **49,** 2275 (1977).

23. Crooks, J. E., D. L. Gerrard, and W. F. Maddams, *Anal. Chem.,* **45,** 1823 (1973).

24. Dawkins, J. V., M. Hemming, *J. Appl. Polym. Sci.,* **19,** 3107 (1975).

25. De Leenheer, A., *J. Chromatogr.,* **74,** 35 (1972).

26. Diaz-Rueda, J., H. J. Sloane, and R. J. Obremski, *Appl. Spectrosc.,* **31,** 298 (1977).

27. Downs, A. J., and S. C. Peake, *Molec. Spectrosc.,* **1,** 523 (1973).

28. Drew, D. M., and J. I. van Gemert, *Appl. Spectrosc.,* **25,** 465 (1971).

29. Edgell, W. F., *Appl. Spectrosc.,* **25,** 276 (1971).

30. Erley, D. S., *Anal. Chem.,* **29,** 1564 (1957).

31. Erley, D. S., B. H. Blake, and A. W. Long, *Appl. Spectrosc.,* **14,** 25 (1960).

32. Fabbri, G., and P. Baraldi, *Appl. Spectrosc.,* **26,** 593 (1972).

33. Fahrenfort, J., *Spectrochim. Acta,* **17,** 698 (1961).

34. Fahrenfort, J., and W. M. Visser, *Spectrochim. Acta,* **18,** 1103 (1962).

35. Fateley, W. G., R. E. Witkowski, and G. L. Carlson, *Appl. Spectrosc.,* **20,** 190 (1966).

36. Feairheller, W. R., and J. T. Miller, Jr., *Appl. Spectrosc.,* **25,** 175 (1971).

37. Ferraro, J. R., and L. J. Basile, *Appl. Spectrosc.,* **28,** 505 (1974).

38. Ford, T. A., P. F. Seto, and M. Falk, *Spectrochim. Acta,* **25A,** 1650 (1969).

39. Freeman, S. K., Gas Chromatography and Infrared and Raman Spectrometry, in L. S. Ettre and W. H. McFadden, Eds., *Ancillary Techniques of Gas Chromatography,* Wiley, New York, 1969, pp. 227–267.

40. Fujiyama, T., and B. Crawford, Jr., *J. Phys. Chem.,* **72,** 2174 (1968).

41. Gallaher, K. L., and J. G. Grasselli, *Appl. Spectrosc.,* **31,** 456 (1977).

42. Garner, H. R., and H. Packer, *Appl. Spectrosc.,* **22,** 122 (1968).

43. Gilby, A. C., J. Cassels, and P. A. Wilks, Jr., *Appl. Spectrosc.,* **24,** 539 (1970).

44. Gomez-Taylor, M. M., D. Kuehl, and P. R. Griffiths, *Appl. Spectrosc.,* **30,** 447 (1976).

45. Gomez-Taylor, M. M., and P. R. Griffiths, *Appl. Spectrosc.*, **31**, 528 (1977).

46. Goulden, J. D. S., *Spectrochim. Acta*, **15**, 657 (1959).

47. Greenberg, J., and L. J. Hallgren, *Rev. Sci. Instrum.*, **31**, 444 (1960).

48. Greenler, R. G., *J. Chem. Phys.*, **44**, 310 (1966).

49. Griffiths, P. R., *Appl. Spectrosc.*, **26**, 73 (1972).

50. Griffiths, P. R., *Appl. Spectrosc.*, **31**, 284 (1977).

51. Griffiths, P. R., and F. Block, *Appl. Spectrosc.*, **27**, 431 (1973).

52. Gruenfeld, M., and R. Ginell, *Appl. Spectrosc.*, **24**, 380 (1970).

53. Haken, J. K., and R. L. Werner, *Appl. Spectrosc.*, **22**, 345 (1968).

54. Hallam, H. E., *Vibrational Spectroscopy of Trapped Species*, Wiley, New York, 1973.

55. Hanst, P. L., A. S. Lefohn, and B. W. Gay, Jr., *Appl. Spectrosc.*, **27**, 188 (1973).

56. Harms, D. L., *Anal. Chem.*, **25**, 1140 (1953).

57. Harrick, N. J., *Anal. Chem.*, **43**, 1533 (1971).

58. Harrick, N. J., *Appl. Spectrosc.*, **31**, 548 (1977).

59. Harrick, N. J., *Internal Reflection Spectroscopy*, Interscience, New York, 1967.

60. Harrick, N. J., and N. H. Riederman, *Spectrochim. Acta*, **21**, 2135 (1965).

61. Hermann, T. S., with S. R. Harvey and C. N. Honts, *Appl. Spectrosc.*, **23**, 435, 451, 461, 473 (1969).

62. Hirschfeld, T., *Appl. Spectrosc.*, **20**, 336 (1966).

63. Hirschfeld, T., *Appl. Spectrosc.*, **21**, 335 (1967).

64. Hirschfeld, T., *Appl. Spectrosc.*, **24**, 277 (1970).

65. Hirschfeld, T., *Appl. Spectrosc.*, **30**, 353 (1976).

66. Hirschfeld, T., *Appl. Spectrosc.*, **31**, 289 (1977).

67. Hordvik, A., *Appl. Opt.*, **16**, 2827 (1977).

68. Jeannotte, A. C., II, and J. Overend, *Spectrochim. Acta*, **33A**, 849 (1977).

69. Johnson, W. T. M., *Offic. Dig. Federat. Soc. Paint Technol.*, **32**, 1067 (1960).

70. Johnson, W. T. M., *Offic. Dig. Federat. Soc. Paint Technol.*, **33**, 1489 (1961).

71. Johnson, W. T. M., and R. Penneys, *Appl. Spectrosc.*, **28**, 328 (1974).

72. Kaczaj, J., *Appl. Spectrosc.*, **21**, 180 (1967).

73. Katlafsky, B., and M. W. Dietrich, *Appl. Spectrosc.*, **29**, 24 (1975).

74. Katon, J. E., *Develop. Appl. Spectrosc.*, **9**, 3 (1971).

75. Katon, J. E., and D. B. Phillips, *Appl. Spectrosc. Rev.*, **7**, 1 (1973).

76. King, S. S. T., *J. Agric. Food Chem.*, **21**, 526 (1973).

77. Kirkland, J. J., *Anal. Chem.*, **29**, 1127 (1957).

78. Lauer, J. L., and M. E. Peterkin, *J. Lubr. Technol.*, **97**, 145 (1975).

79. Lauer, J. L., and M. E. Peterkin, *J. Lubr. Technol.*, **98**, 230 (1976).

80. Littlewood, B., *Chromatographia*, **1**, 223 (1968).

81. Louw, C. W., and J. F. Richards, *Appl. Spectrosc.*, **29**, 15 (1975).

82. McCall, E. R., N. M. Morris, V. W. Tripp, and R. T. O'Connor, *Appl. Spectrosc.*, **25**, 196 (1971).

83. McNiven, N. L., and R. Court, *Appl. Spectrosc.*, **24**, 296 (1970).

84. Milligan, D. E., and M. E. Jacox, Matrix Spectra, in K. N. Rao and C. W. Mathews, Eds., *Molecular Spectroscopy: Modern Research*, Academic Press, New York, 1972.

85. Milne, J. W., *Spectrochim. Acta*, **32A**, 1347 (1976).

86. Mitzner, B. M., *J. Opt. Soc. Am.*, **47**, 328 (1957).

87. Nadeau, A., and R. N. Jones, *Spectrochim. Acta*, **26A**, 742 (1970).

88. Norris, W. P., A. L. Olsen, and R. G. Brophy, *Appl. Spectrosc.*, **26**, 247 (1972).

89. Nortia, T., and E. Kontas, *Spectrochim. Acta*, **29A**, 1493 (1973).

90. O'Connor, R. T., Cellulose and Fabrics, in G. L. Clark, Ed., *Encyclopedia of Spectroscopy*, Reinhold, New York, 1960, pp. 392–409.

91. O'Connor, R. T., E. F. Du Pré, and E. R. McCall, *Anal. Chem.*, **29**, 998 (1957).

92. Paralusz, C. M., *J. Colloid Interface Sci.*, **47**, 719 (1974).

93. Parliment, T. H., *Microchem. J.*, **20**, 492 (1975).

94. Percival, C. J., and P. R. Griffiths, *Anal. Chem.*, **47**, 154 (1975).

95. Perenich, T. A., and E. C. Tuazon, *Appl. Spectrosc.*, **30**, 196 (1976).

96. Perry, J. A., *Appl. Spectrosc. Rev.*, **3**, 229 (1970).

97. Pitts, J. N., B. J. Finlayson-Pitts, and A. M. Winer, *Environ. Sci. Technol.*, **11**, 568 (1977).

98. Poling, G. W., *J. Electrochem. Soc.*, **116**, 958 (1969).

99. Potts, W. J., Jr., *Chemical Infrared Spectroscopy*, Vol. 1, *Techniques*, Wiley, New York, 1963.

100. Potts, W. J., Jr., and N. Wright, *Anal. Chem.*, **28**, 1255 (1956).

101. Price, W. C., and K. S. Tetlow, *J. Chem. Phys.*, **16**, 1157 (1948).

102. Rochkind, M. M., *Anal. Chem.*, **40**, 762 (1968).

103. Rochkind, M. M., *Spectrochim. Acta*, **27A**, 547 (1971).

104. Ruppin, R., and R. Englman, *Rep. Progr. Phys.*, **33**, 149 (1970).

105. Schiedt, U., and H. Reinwein, *Z. Naturforsch.*, **7b**, 270 (1952).

106. Schwarz, H. P., R. C. Childs, L. Dreisbach, S. V. Mastrangelo, and A. Kleschick, *Appl. Spectrosc.*, **12**, 35 (1958).

107. Shaps, R. H., and A. Varano, *Industr. Res.*, **19**, (2), 86 (1977).

108. Smith, A. L., Infrared Spectroscopy, in J. W. Robinson, Ed., *Handbook of Spectroscopy*, Vol. II, CRC Press, Cleveland, 1974.

109. Smith, D. C., and E. C. Miller, *J. Opt. Soc. Am.*, **34**, 130 (1944).

110. Staats, P. A., and H. W. Morgan, *Appl. Spectrosc.*, **22**, 576 (1968).

111. Stimson, M. M., and M. J. O'Donnell, *J. Am. Chem. Soc.*, **74**, 1805 (1952).

112. Strait, L. A., and M. K. Hrenoff, A Technique for Preparing Solid Organic Samples for Infrared Analysis, in *Symposium on Spectroscopy*, ASTM Spec. Tech. Publ. No. 269, 190, 1960.

113. Strasheim, A., and K. Buijs, *Spectrochim. Acta*, **16**, 1010 (1960).

114. Sumas, E. C., J. F. Williams, C. Walker, and D. Kidd, *Appl. Spectrosc.*, **27**, 486 (1973).

115. Thomas, G. J., Jr., and Y. Kyogoku, Biological Science, in E. G. Brame, Jr., and J. G. Grasselli, Eds., *Practical Spectroscopy*, Vol. 1, *Infrared and Raman Spectroscopy*, Part C, p. 717; Marcel Dekker, New York, 1977.

116. Tolk, A., *Spectrochim. Acta*, **17**, 511 (1961).

117. Tompa, A. S., *Appl. Spectrosc.*, **22**, 491 (1968).

118. Tompkins, H. G., *Appl. Spectrosc.*, **28**, 335 (1974).

119. Tompkins, H. G., *Appl. Spectrosc.*, **30**, 377 (1966).

120. Truett, W. L., *Am. Lab.*, **9** (6), 33 (1977).

121. Wall, D. L., and A. W. Mantz, *Appl. Spectrosc.*, **31**, 552 (1977).

122. Watson, C. A., *Anal. Chem.*, **49**, 835A (1977).

123. Welti, D., *Infrared Vapour Spectra*, Heyden, London, 1970.

124. White, J. U., *J. Opt. Soc. Am.*, **32**, 285 (1942).

125. White, J. U., *Rev. Sci. Instrum.*, **21**, 629 (1950).

126. Wilks, P. A., Jr., *Appl. Spectrosc.*, **22**, 782 (1968).

127. Wilks, P. A., Jr., and T. Hirschfeld, *Appl. Spectrosc. Rev.*, **1**, 99 (1967).

128. Willey, R. R., *Appl. Spectrosc.*, **30**, 593 (1976).

QUALITATIVE APPLICATIONS

INTRODUCTION

The IR spectrum of an organic substance has often been called its "fingerprint." Since the spectrum of each molecular species is unique, IR is a powerful tool for the identification of organic compounds.

Infrared is also extremely useful in the elucidation of unknown structures. This is true because of the carryover of "group frequencies" from one compound to another. That is, certain chemical groups (methyl, carbonyl, etc.) give characteristic absorption bands that are more or less constant in wavelength and intensity. Group frequencies are discussed in more detail after a brief introduction to molecular physics.

THEORY OF IR ABSORPTION

In the general case a molecule can have the following types of motions: (1) translation of the whole molecule, which can be regarded as a translation of the center of mass, (2) rotation of the molecule as a framework around its center of mass, (3) vibrations of the individual atoms within the framework, which occur in such a manner that the position of the center of mass is not changed and the framework does not rotate, (4) motions of the electrons inside the molecule, and (5) spins of the electrons and the nuclei of the atoms. Situations (1), (4), and (5) are not of direct concern in analytical infrared spectroscopy and are not considered further.

The number of *degrees of freedom* (DF) of a particle equals the number of coordinates required to specify its position in space. The total DF of a system of N particles is equal to $3N$. If a molecule is considered as a rigid framework, we can determine its position in space by specifying the three coordinates of its center of mass. But the molecule also has three rotational DF, so three additional coordinates are needed to fix its orientation. Therefore, the number of vibrational DF is equal to $3N - 3 - 3$ or $3N - 6$. (Linear molecules, such as HCl, O_2, CO_2, and C_2H_2, have only two rotational DF since rotation in the usual sense does not occur around the figure axis of the molecule. Thus a linear molecule has $3N -$

5 degrees of vibrational freedom.) This means that there are exactly $3N - 6$ fundamental frequency modes with which a nonlinear molecule of N particles can vibrate.

The ability of a compound to absorb IR energy depends on a net change in dipole moment occurring when the molecule rotates or vibrates. Whether or not such a change occurs depends on the distribution of electrical charges in the molecule. As an example, the nitric oxide (NO) molecule can be thought of as a nitrogen atom joined to an oxygen atom by means of a compressible bond. The oxygen has eight electrons surrounding its nucleus and the nitrogen has seven. During a vibration, a change in the charge distribution occurs; that is, the radiation beam sees an oscillating charge. Consequently, nitric oxide shows an absorption band in the IR region at the frequency corresponding to the vibrational frequency of the nuclei. On the other hand, there is no net change in the dipole moment during the vibration of homonuclear molecules such as O_2, H_2, N_2, or Cl_2, and these gases do not absorb IR radiation. A little thought will show that the ability of a molecule to absorb radiation during a particular vibration depends only on its electrical geometry. Mathematical methods have been worked out for predicting the IR activity of the various vibrations of a molecule on the basis of its symmetry [119].

The total energy of a molecule is given (to a good degree of approximation) as the sum of its translation, rotational, vibrational, and electronic energies. That is

$$E_{total} = E_{trans} + E_{rot} + E_{vib} + E_{elect} \qquad (5.1)$$

Since the translational energy has little effect on molecular spectra, it is disregarded here.

It was first suggested by Niels Bohr in 1913 that atoms and molecules cannot possess any arbitrary energy, but must exist in discrete energy states. Consequently, transitions between energy states result in absorption or emission of characteristic units of energy (quanta), which are observed variously as emission lines from excited molecules or as absorption bands in the IR, visible, and UV regions. All spectroscopy is based on this important concept.

The subsequent development of quantum mechanics gave a more accurate method of describing atomic phenomena than was available from classical mechanics. The semiclassical picture presented below is perhaps easier to visualize, however.

Rotational Spectra

Rotating molecules *in the gas or vapor state* are restricted to definite energy states; therefore, they absorb radiation of frequencies corresponding to transitions between rotational energy states. Since the energies involved are low, rotational absorption bands usually fall in the far-IR region (beyond 500 cm⁻¹).

Molecules may be classified according to their rotator symmetry. The simplest molecules giving rotational spectra are the diatomic ones, called *dumbbell* rotators. The next simplest molecules from the standpoint of rotator symmetry are the spherical tops (three equal moments of inertia), such as methane.

Spherical tops do not show rotational IR spectra because there is no change in dipole moment as the molecule rotates. A more complicated type is the *symmetrical top* (two equal moments of inertia), examples of which are NH_3 and C_6H_6. The most complicated rotators are the *asymmetrical tops* (all three moments unequal), such as H_2O, C_2H_4, and most polyatomic molecules.

The rotational energy, E_{rot}, depends on the rotator symmetry and the moments of inertia of the molecule being considered. It also depends on the rotational quantum number, J, which is related to the angular momentum of the molecule by the expression

$$\text{Angular momentum} = \sqrt{J(J + 1)}\,\frac{h}{2\pi} \tag{5.2}$$

where J takes on integral values, just as do most other quantum numbers. For rigid dumbbell and spherical top rotators,

$$E_{rot} = J(J + 1)Bhc \tag{5.3}$$

where $B = h/(8\pi^2 cI)$; h is Planck's constant; c is the speed of light; and I is the moment of inertia about the axis of rotation. For the transition $J = 0 \to J = 1$, the energy change is

$$\Delta E = 2Bhc \tag{5.4}$$

Equation (5.3) tells us that the larger the moment of inertia of a molecule, the more closely the rotational lines are spaced. A typical rotational-level diagram is shown in Fig. 5.1, along with the rotational spectrum of a diatomic molecule. Selection rules are $\Delta J = \pm 1$ for IR transitions.

Vibrational Spectra

To understand the behavior of a vibrating molecule, we first consider a simple vibrating system—a weight on a spring. The kinetic or "moving" energy K is given by the expression

$$K = \tfrac{1}{2}\,mv^2 \tag{5.5}$$

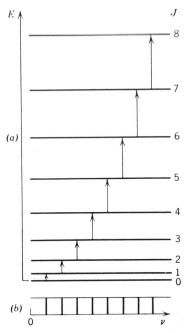

Fig. 5.1   Energy levels and transitions of rigid rotator (*a*) and resulting rotational spectrum (*b*). *Reproduced, with permission of Van Nostrand-Reinhold, from G. Herzberg, Spectra of Diatomic Molecules,* Princeton, N. J., 1950.

where m is the mass and v is the velocity of the weight. The potential energy, V, due to the elongation or compression of the spring, has its largest value when the weight is at the bottom or the top of its cycle. (When the spring is stretched or compressed, it has the ability to do work and thus possesses potential energy.) The sum of the kinetic and potential energies is constant. If x is the displacement of the spring from its equilibrium position, force is required to produce that displacement, and the restoring force F is proportional to the displacement but in the opposite direction; that is,

$$F = -kx \tag{5.6}$$

where k is the proportionality factor or "force constant." Now force is the negative derivative of the potential energy, or

$$F_x = -\frac{\partial V}{\partial x} = -kx \tag{5.7}$$

Integrating, we obtain

$$V = \int_0^x kx \ dx = \tfrac{1}{2} kx^2 \tag{5.8}$$

which is the expression for the potential energy. An oscillator with a potential function such as is described in equation (5.8) is known as a "simple harmonic oscillator." The relation of the kinetic, potential, and total energies is shown in Fig. 5.2.

To solve for the frequency of the oscillation, we use Newton's second law of motion, which states that force equals mass times acceleration, or

$$F = ma \tag{5.9}$$

However, the acceleration is the second derivative of distance (x) with time, so

$$F = m\frac{\partial^2 x}{\partial t^2} = m\ddot{x} \tag{5.10}$$

Combining equations (5.7) and (5.10), we have

$$\ddot{x} = -\frac{k}{m}x \tag{5.11}$$

The solution, x, must be a repeating function such that its second derivative is equal to the original function times ($-k/m$). A cosine is such a

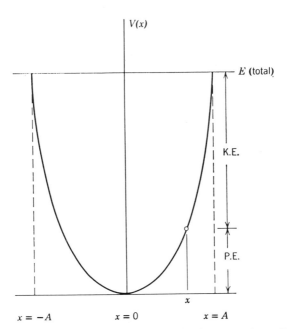

Fig. 5.2 Relationship of kinetic and potential energies for harmonic oscillator with total energy E.

function, and the instantaneous displacement of the oscillator at time t is

$$x = A \cos (2\pi ft + b) \tag{5.12}$$

where f is the frequency of vibration; A is its maximum amplitude; and b is a phase constant that can be made zero by proper choice of boundary conditions. Then

$$\ddot{x} = -4\pi^2 f^2 A \cos (2\pi ft) = -4\pi^2 f^2 x \tag{5.13}$$

Now $\ddot{x} = -(k/m)x$ if

$$k/m = 4\pi^2 f^2 \tag{5.14}$$

so the frequency of the oscillator is

$$f = \frac{1}{2\pi} \sqrt{\frac{k}{m}} \tag{5.15}$$

This equation holds approximately for molecular systems, and in the case of a diatomic molecule the frequency (vibrations per second) is given as

$$f = \frac{1}{2\pi} \sqrt{\frac{k}{\mu}} = \frac{1}{2\pi} \sqrt{k \frac{m + M}{mM}} \tag{5.16}$$

where M and m are the masses of the two atoms and μ is their reduced mass.

In polyatomic molecules the expression is much more complicated because each vibration usually involves many atoms and thus many force constants.

Another complication arises even in the case of diatomic molecules, because actual molecular vibrations are not "simple harmonic." This means that the potential energy may not be symmetrical about the equilibrium position of the vibrating mass, and in any case is not given by a simple expression such as equation (5.8). Instead, V may be written:

$$V = \tfrac{1}{2} kx^2 + Ax^3 + Bx^4 + Cx^5 + \cdots + \tag{5.17}$$

If the anharmonicity is small, the displacement may be represented by

$$x = A_1 \cos (2\pi ft + b_1) + A_2 \cos (2\pi 2ft + b_2) +$$
$$A_3 \cos (2\pi 3ft + b_3) + \cdots + \tag{5.18}$$

For molecules, the potential energy curves are similar to the one shown in Fig. 5.3.

As in the case of rotational energies, vibrational energies are quantized. This means that only specific energy levels can be occupied by an oscil-

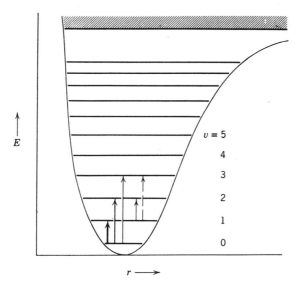

Fig. 5.3 Potential-energy curve for anharmonic oscillator, showing some possible transitions.

lator. In its lowest energy state the molecule is not at rest but vibrates with a certain zero-point energy. The energy of a harmonic oscillator in quantum mechanics is

$$E_{vib} = (v + \tfrac{1}{2})\, hc\nu \tag{5.19}$$

Here v is the vibrational quantum number, which can take on only nonnegative integral values, and ν is the frequency of the fundamental vibration in cm^{-1}. The selection rules are $\Delta v \pm 1$ so the spectrum of such an oscillator consists of only a single line, since the energy levels are all spaced equally. The effect of anharmonicity is to: (1) weaken the selection rules, so transitions of $\Delta v = \pm 2, \pm 3, \ldots$, can occur (overtones will appear) and (2) change the spacing of the higher energy levels (usually to put them closer together). Although both mechanical and electrical anharmonicity are present in an actual molecule, only electrical anharmonicity is required for the appearance of overtones. The population of molecules in the various vibrational levels is determined by the Boltzmann factor.

Some vibrations may be *degenerate*, that is, their frequencies will be the same. Degeneracy of vibration ν_2 is illustrated in Fig. 5.4 for carbon dioxide. In this case fewer than $(3N - 6)$ or $(3N - 5)$ frequencies may appear in the spectrum. *Accidental degeneracy* occurs when two different

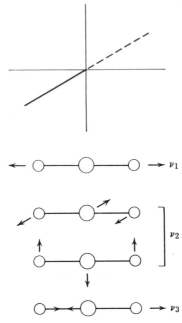

Fig. 5.4 Vibrations of CO_2.

vibrations have nearly the same energy. In this case, the frequencies sometimes interact to repel each other, and one vibration may borrow intensity from the other [119, p. 215].

Although pure IR rotational spectra for most molecules are observed only in the far IR, rotational energy levels are superimposed on vibrational states (Fig. 5.5), so that gas or vapor spectra invariably show rotational "wings" on the main absorption. The lowest-frequency lines (transition $\Delta J = -1$) are called the *P-branch*; absorption at the vibrational frequency is known as the *Q-branch* ($\Delta J = 0$); and the higher-frequency lines ($\Delta J = +1$) make up the *R-branch*. Some vibrations do not show a *Q*-branch.

The origin of a particular vibrational band can sometimes be inferred from the contours of the vibration–rotation envelope (see p. 204). Interactions between rotation and vibrational energy levels caused by centrifugal stretching and Coriolis perturbations (particularly in degenerate vibrations) may distort the band shape.

In the vibrations of polyatomic molecules, all the atoms in the framework move together in phase but with different amplitudes. Thus, strictly speaking, no vibration is localized within a particular group, although it

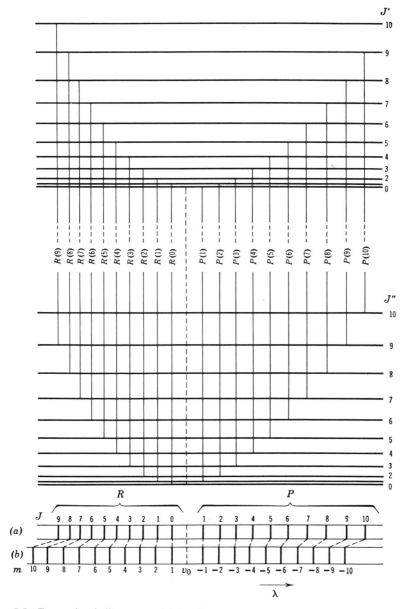

Fig. 5.5 Energy-level diagram explaining fine structure of rotation–vibration band and resulting spectra (*a*) with and (*b*) without allowance for interaction between rotation and vibration. Shorter wavelengths are at left. Reproduced, with permission of Van Nostrand-Reinhold, from G. Herzberg, *Spectra of Diatomic Molecules*, Princeton, N. J., 1950.

may be very nearly so. To illustrate, we note some of the vibrations of $H_2C{=}CBr_2$ in Fig. 5.6 [229]. It can be seen that the CH stretch (3028 cm^{-1}) has a $C{=}C$ stretch component, as does the CH bend (1372 cm^{-1}). This mechanical effect in vibrating systems is one reason that group frequencies change from one molecule to another even if force constants do not (see pp. 144–148). It is obvious that description of a mode as "CH stretch" or "NH bend" is approximate at best, and actual motions may be much more complex than is implied by their descriptive names. Occasionally vibrations are so badly mixed or delocalized that no single term is adequate to describe the motion.

Normal Modes

Each of the $3N - 6$ vibrations of a polyatomic molecule is called a *normal mode* ("normal" is used in the sense of being independent of the other modes). The form of the normal mode is easy to deduce for a diatomic molecule; for polyatomic molecules, derivation of the normal modes is much more difficult since it may involve solution of $3N - 6$ or

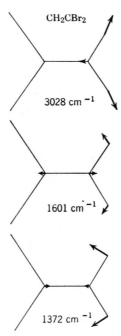

Fig. 5.6 Forms of several vibrations in CH_2CBr_2. Reproduced, with permission, from *Journal of Chemical Physics* [229].

more simultaneous equations involving masses and force constants. Furthermore, there are usually more force constants than frequencies. Nevertheless, by using additional information such as the symmetry properties of the molecule, force constants transferred from similar molecules, frequencies from isotopically substituted molecules, and by making certain simplifying assumptions, one can often make a successful (if laborious) attack on the problem. The use of electronic computers has greatly expedited such calculations and as a result, normal coordinate analyses are now being applied to molecules of considerable complexity.

It is important to note that the symmetry of the vibration determines its infrared and Raman activity. By methods of group theory [61, 88] tables have been constructed with which one can predict the number and activity of vibrational modes of each symmetry class.

The method used to classify vibrations by symmetry and the mathematical formulation of the normal coordinates is beyond the scope of this chapter. A more complete qualitative description is developed by Potts [215], and rigorous treatments are given by Herzberg [119] and by Wilson, Decius, et al. [272].

The first step in deriving the normal coordinates is to set up a potential function. In writing the potential function, one ordinarily uses internal rather than Cartesian coordinates. Internal coordinates are usually expressed in terms of changes in interatomic distances and angles. They have the advantage of giving a more significant picture of the molecular vibrations, and expedite the algebraic operations.

The numerical values obtained for the force constants will depend on the form of the potential function used to set up the equations of motion. When comparing force constants, therefore, one should make certain that they were derived using the same type of force field. A completely general potential function would include interaction terms between all atoms of a molecule, whether bonded or not. Such a function is unsatisfactory, first because the force constants have little, if any, physical meaning and second, because the large number of unknown terms makes their solution impossible for any but the very simplest molecules.

More meaningful functions have been devised. The valence-bond force field assumes that the significant forces involved in a molecular vibration occur along chemical bonds and, therefore, expresses the force constants in terms of changes in bond lengths and angles. The central force field approach treats the molecule without regard to the valence bonds and defines the force constants in terms of interactions between atoms, whether bonded or not. Since neither of these force fields is completely satisfactory, a modified force field, known as the Urey–Bradley field (also unsatisfactory at times!) has been proposed. The Urey–Bradley field is essentially a combination of the valence bond and central force fields. It assumes that the important forces act along chemical bonds but also includes interaction terms between *adjacent* nonbonded atoms. Such a force field can thus take into account attractions and repulsions between adjacent nonbonded atoms (field effects), which have been found to have some influence on the form, intensity, and frequency of normal vibrations. Furthermore, if the coordinates are chosen carefully, physically significant force constants may be derived that can be transferred among molecules of similar structure.

Once the force constants have been determined, the form of each vibration may be derived. For each frequency, there exists a single normal coordinate that describes the displacements of all the atoms involved in that vibration. The normal coordinate, by a suitable transformation, may be expressed in terms of Cartesian or valence coordinates, which give the actual displacements of the atoms in an easily visualized form. Details of this and other operations involved in normal coordinate calculations are given elsewhere [56, 78, 244].

Each of the $3N - 6$ normal modes may be represented by its own potential energy diagram, similar to Fig. 5.3. The abscissa is then the normal coordinate rather than the interatomic distance.

A short explanation of the conventions used in designating symmetry classes of vibrations and in numbering fundamentals may be useful. In the spectroscopic literature, vibrations that are symmetric with respect to an axis of symmetry are designated as type A. Subscripts or primes indicate subclasses within the species. Antisymmetric vibrations are classed as type B, doubly degenerate ones by E, and triply degenerate modes by F. Individual vibrations are numbered starting with the highest-frequency, totally symmetric mode and numbering through that class in order of decreasing frequency; then proceeding to the next most symmetric mode and again numbering in order of decreasing frequency. Thus for $HCCl_3$, the CH stretch (Class A_1), $\nu_1 = 3033$ cm^{-1}; CCl_3 symmetrical stretch (Class A_1), $\nu_2 = 667$ cm^{-1}; CCl_3 symmetrical deformation (Class A_1), $\nu_3 = 364$ cm^{-1}; CH bend (Class E), $\nu_4 = 1205$ cm^{-1}; CCl_3 asymmetric stretch (Class E), $\nu_5 = 760$ cm^{-1}; and CCl_3 asymmetrical deformation (Class E), $\nu_6 = 260$ cm^{-1} (descriptions are approximate). Further discussion and numerous examples of these conventions are given by Herzberg [119].

Certain descriptive terms are often used in vibrational spectroscopy, (e.g., CH_2 rock; CCl_3 torsion; or CF_2 wag). Some of these motions are depicted in Fig. 5.7; the exact form of the modes depends on the masses and force constants of the atoms involved. Symbols often used as a

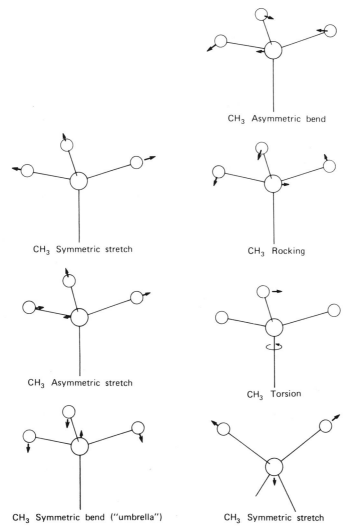

CH₃ Asymmetric bend

CH₃ Symmetric stretch

CH₃ Rocking

CH₃ Asymmetric stretch

CH₃ Torsion

CH₃ Symmetric bend ("umbrella")

CH₃ Symmetric stretch

Fig. 5.7 Approximate forms of normal modes for some common organic groups. Reproduced, with permission, from Smith [242]. Copyright The Chemical Rubber Company, CRC Press, Inc.

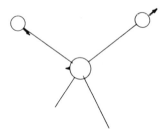

CH₂ Asymmetric stretch.

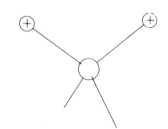

CH₂ Wag.

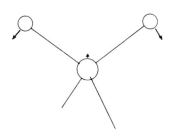

CH₂ Symmetric bend ("scissors").

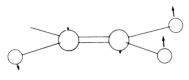

Vinyl CH₂ wag.

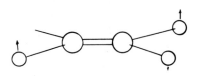

Vinyl C=C twist.

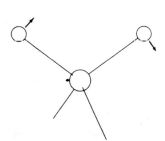

CH₂ Rock.

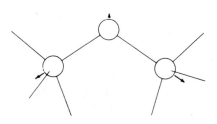

COC Symmetric stretch.

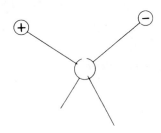

CH₂ Twist.

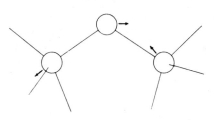

COC Asymmetric stretch.

Fig. 5.7 (Continued)

136

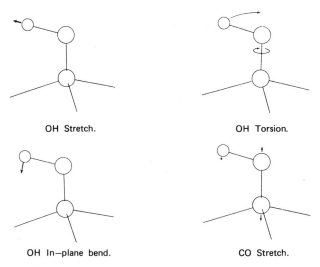

OH Stretch.

OH Torsion.

OH In—plane bend.

CO Stretch.

Fig. 5.7 (Continued)

shorthand description for these motions include ν (stretch); δ (deformation); ρ (rocking); ω (wag); and τ (torsion).

Band Shapes

We have noted that the vibrational absorptions in vapors and gases are modified by the superposition of rotational energy levels. In the liquid or solution state, the rotational wings disappear because rotation is highly restricted. (Molecules with small moments of inertia dissolved in nonpolar solvents seem to show unquantized rotation [146]). Rather than sharp lines, all absorption bands have contours that are symmetrical about a central maximum, with low-intensity wings extending some distance to each side. Factors contributing to this distribution of intensities in gases [223] include natural line breadth due to radiation damping, the Doppler effect, pressure broadening, and specific intermolecular interactions. In condensed phases, collisions by nearest neighbors and specific interactions are chiefly responsible for band shapes. Isotopic splitting, Fermi resonance, and "hot" bands (see p. 139) may also at times be important.

A number of mathematical formulations have been proposed to describe empirically observed band contours. One of the most successful of these is the Lorentz function, which takes the form

$$\log_e \frac{I_0}{I} = \frac{a}{(v - v_0)^2 + b^2} \qquad (5.20)$$

where $(a/b^2) = \log (I_0/I)_{max}$; $2b = \Delta v_{1/2} =$ band width at half height; I_0 = band intensity at maximum; $v_0 =$ frequency at maximum.

The Lorentz function is useful in calculating band contours for resolution of complex absorption shapes in the liquid and in calculation of integrated intensities. Intrinsic and instrumental factors influencing band shapes, methods of correcting observed shapes to obtain true profiles, and the determination of integrated intensities have been reviewed by Seshadri and Jones [233]. A "pseudodeconvolution" calculation for removing instrumental distortions from observed band shapes has been described [145].

Band Intensities

The question of IR band intensities is so intimately tied up with changes in electron configuration during vibration that it apparently can be resolved only by calculating the total wave function for the molecule. Mathematical descriptions of the phenomenon have been formulated and are of some value in calculating bond moments, but *a priori* calculation of absorption intensities are of such crudity as to be almost without value. Empirical correlations of intensity with structure are sometimes useful, however.

If a_v is the absorption coefficient at frequency v, the basic relationship between observed intensities and bond moment μ is (in the first approximation)

$$\int a_v \, dv = \frac{N \pi}{3c^2} \left[\frac{\partial \mu}{\partial Q} \right]^2 \qquad (5.21)$$

where Q is the normal coordinate for the vibration, N is the number of molecules per cubic centimeter of sample, and c and π have their usual meaning. The left-hand side of equation (5.21) can be measured accurately for isolated bands if a narrow spectral slit width is used; for finite slit widths, correction procedures must be applied [143, 224]. Even after an accurate integral is obtained (on a frequency vs. absorbance—*not* percent transmission—plot) there remains an ambiguity in sign resulting from the square-root operation. Interpretation of the right-hand side of equation (5.21) has been considered by Coulson [62], who points out the problems involved in understanding the bond moment and its rate of change in Q. Further discussion of band intensities may be found elsewhere [119, pp. 260 ff.; 233; 272, pp. 162 ff.].

Overtone and Combination Bands

Anharmonicity weakens the selection rules for harmonic oscillators so that overtone and combination bands are allowed (Fig. 5.3). In general, the overtone absorptions of a vibration ν_1 will fall at approximately $(2\nu_1)$ $- b$, where b is 2–10 cm^{-1} (occasionally more). Some bands exhibit negative anharmonicity; that is, the overtone falls at a frequency greater than twice that of the fundamental. It is also possible that, if the fundamental vibration has the proper symmetry, the first overtone may not appear at all. Overtone activity can be predicted from the symmetry of the vibration [119, pp. 126 ff.]. Combination bands similarly fall at a frequency slightly less than the sum of the fundamental frequencies. Difference bands, which originate from an excited vibrational state, are sometimes seen. In this case, however, the relationship $\nu_{12} = (\nu_1 - \nu_2)$ holds exactly. If a difference band appears in the spectrum, the corresponding summation band should also appear. Another kind of difference band, commonly known as a "hot" or "upper-stage" band, also involves a transition from an excited state to another excited level. The frequency may be written $\nu = (\nu_j + \nu_k) - \nu_j$, or for the transition $(v = 1) \rightarrow (v = 2)$, $\nu = 2\nu_k - \nu_k$, and it will be a few wave numbers different from ν_k. Hot bands are usually weak and seldom observed except in high-resolution spectroscopy of gases.

The intensity of overtone and combination bands in condensed states is usually less than that of the fundamental by a factor of 10–100. Exceptions occur; the first overtone of the —NCS bending frequency, for example, characteristically has about the same intensity as the fundamental.

Bands observed in the near-IR region (0.7–2.5 μ) are, with the exception of a few electronic transitions, almost all overtones or combinations of hydrogen stretching vibrations or combinations of hydrogen stretches with other vibrations. There are two reasons for this. First, since the mass of hydrogen is small, it makes large excursions during a vibration. Consequently, the motion has considerable anharmonicity, which leads to a greater overtone band intensity. Second, most nonhydrogenic vibrations lie at lower frequencies so that only second and higher overtones and multiple combinations (which are, in turn, much weaker than first harmonics) fall in this region of the spectrum. Most near-IR work, therefore, is done using the numerous overtone and combination frequencies from hydrogen vibrations. In some cases, anharmonicity constants for different vibrations are sufficiently different that bands superimposed in the IR region are resolved in the near IR. A short discussion of spectroscopy in the near-IR region, together with a bibliography of methods and applications (to 1958), is given by Kaye [154]. A review covering this region may be useful [266]. Correlation charts for the near IR have also been published [56, 102].

GROUP FREQUENCIES: USES AND LIMITATIONS

General

Certain chemical groups such as CH_3, $C{=}O$, $P{=}O$, and $C{=}C$ have been found empirically to absorb at very nearly the same wavelengths regardless of the molecule in which they are found. Such absorptions are called "group frequencies" and often represent a rapid, unambiguous means of confirming the presence or absence of the chemical moiety responsible for the absorption. Rarely, if ever, do we find a completely invariant group frequency, however. Rather, the position of the band varies within a range of frequencies that is defined by both internal and external influences on the vibrating molecule. In this section we briefly consider internal effects (electronic and mechanical) on the vibrating group; external (environmental) influences on the spectra are discussed in the following section.

We recall that the vibration frequency of a two-body system is

$$f = \frac{1}{2\pi} \sqrt{\frac{k}{\mu}} \tag{5.22}$$

where f is the vibrational frequency, k is the force constant of the bond, and μ is the reduced mass. It is easy to calculate μ for a diatomic molecule, but k can be determined accurately only by experiment.

In an effort to develop an empirical relationship between the force constant for vibrating diatomic molecules and other physical parameters, several investigators have studied the vibrational spectra of numerous diatomic species. An expression by Badger [9] that has found considerable use is

$$k_0 = a_{ij} (D_e - b_{ij})^{-3} \tag{5.23}$$

where k_0 is the force constant (dynes/cm); D_e is the equilibrium interatomic distance, and a_{ij} and b_{ij} are constants that depend on the rows in the periodic table in which atoms i and j are found.

A more refined relationship, proposed by Gordy [106], has the form

$$k = 1.67N \left[\frac{X_A X_B}{d^2} \right]^{3/4} + 0.30 \qquad (5.24)$$

where k is the force constant; N is the bond order; d is the bond length; and X_A and X_B are the electronegativities of atoms A and B, respectively.

The important points about these and similar expressions are: (1) results are only approximate, since the stretching force constant depends on the shape of the potential energy curve, which is a function of electron configuration as well as of the position of the neighboring electronic states; and (2) strictly speaking, these formulas are applicable only to diatomic molecules, although they can be used with polyatomic molecules if the vibration is localized between two groups in such a way as to simulate a diatomic molecule. Prudent usage of these functions suggests that they be used only to calculate approximate frequencies and that specific band assignments be based on firmer evidence.

We now discuss the kinds of vibrations responsible for the appearance of group frequencies in the spectrum. It has been noted previously that a molecule can be visualized as a set of N masses connected by springs (chemical bonds), and that the vibrations of this system can be resolved into $3N - 6$ normal modes. Strictly speaking, any of these normal modes involves motion of all N atoms. To a very good approximation, however, some of these vibrations are localized in a particular group, and it is such vibrations that give rise to group frequencies.

One such case occurs when we have a light atom, such as hydrogen, vibrating against a heavier atom, such as carbon, as in a C—H stretching vibration. The motion is largely that of the lighter atom and consequently is not affected very much by the rest of the molecule.

A second type of group frequency results when the atoms involved are of similar masses, but the vibration couples only weakly to the rest of the molecule. Examples of this situation are found in the multiple-bond frequencies such as >C=O and C≡N stretching modes. Similarly, more or less isolated groups such as a phenyl group on carbon or a methyl group on silicon have a low degree of vibrational coupling to the remainder of the molecule for at least some of their normal frequencies.

These types of group absorptions are almost always recognizable because they fall within relatively narrow wavelength limits. A class of group frequencies also exists in which the absorptions may vary over a considerable range in wavelength, but the position of the band can be correlated with mass, resonance, or electronic influences from the remainder of the molecule. At the present time, the reasons for some of these correlations are poorly understood, and they should be considered largely as empirical (but useful) relationships.

Thus far we have stressed the *position* of the bands on the frequency scale as related to molecular structure. Frequency measurements can be made accurately and transferred readily between different instruments. Perhaps equally important, but certainly less well understood, are variations of *intensity, band shape,* and *width*. These parameters are used instinctively by every spectroscopist, but only recently has research effort been focused on the problem of their quantitative comparison.

Widespread use of the intensities for qualitative analysis predicted several years ago has not yet materialized, partly because of the difficulty of separating overlapping bands of a complex spectrum and partly because most spectrometers do not permit accurate integrated intensities to be obtained without considerable effort. The problem of quantitatively reconciling band shapes with known molecular parameters has received even less attention. Here again, instrumental limitations and band overlap have tended to deter research in this area.

Internal Factors Influencing Group Frequencies

If chemical bonds were always the same regardless of the compound in which they were found, problems in molecular dynamics could be solved easily by the methods of classical mechanics. As every chemist knows, however, many subtle factors influence the length, polarity, direction, and strength of the bonds. Factors that directly influence group frequencies and/or intensities in molecular spectra include changes in atomic mass, vibrational coupling, resonance, inductive and field effects, conjugation, hydrogen bonding, and bond-angle strain [56]. These perturbing influences are discussed in the following sections.

Mass Changes

From equation (5.16) relating mass and frequency, we would expect a stretching frequency between atoms A and B, on substitution of C for B, to change in the ratio of the square root of the reduced masses of B and C. That is,

$$\frac{f_{AB}}{f_{AC}} = \sqrt{\frac{\mu_{AC}}{\mu_{AB}}} \tag{5.25}$$

with other things being equal. Except in the case of isotopic substitution, however, other things are rarely equal, because replacement of one atomic species by another alters bond polarity, length, and strength as well as mass. If, for example, we calculate the expected frequencies of the hydrogen halides using HI as standard (i.e., assume the force constant for the other hydrogen halides to be the same as for HI), we find for HBr, 2314 cm^{-1} (calculated) and 2564 cm^{-1} (observed); for HCl, 2333 cm^{-1} (calculated), and 2886 cm^{-1} (observed); and for HF, 2360 cm^{-1} (calculated), and 2968 cm^{-1} (observed). Obviously, in this series, force constants change as well as masses. The bond length also changes with the size of the halogen atom, and infrared frequencies are a sensitive measure of bond length [191, 193].

If A, B, or C represents a group of atoms, the simple expression given in equation (5.25) no longer applies because the effective mass of the group is not the sum of the atomic masses. This is true because, although all the atoms of the group vibrate in phase, they do not all vibrate in the same direction with the same amplitude. If we compare the C—Cl stretch vibration in CH_3Cl and CH_3CH_2Cl, for example (where electrical effects should be approximately equivalent), by assuming equal force constants and masses of 15 and 29 for the CH_3 and C_2H_5 groups, respectively, we calculate a C—Cl frequency of 608 cm^{-1} for CH_3CH_2Cl compared to 657 cm^{-1} observed. This result shows that the effective mass of the ethyl group is less than the sum of its atomic masses, as we might have anticipated intuitively. Thus where groups of atoms are concerned, we cannot usually relate frequency changes directly to mass without a normal coordinate analysis, even where electrical and steric effects are similar.

Isotopic substitution usually results in a stretching frequency very near that predicted. The replacement of hydrogen by deuterium is a well-known technique for studying vibrations involving hydrogen. For a pure stretching mode where D is substituted for H vibrating against an infinite mass, we expect the frequency to shift by the factor $\sqrt{2}$ or 1.414, but in an actual molecule the shift depends on the vibrational mode involved. Krimm [161, 162] has shown, for example, that for deuterium substitution in an isolated CH_2 group the frequency shifts for the different modes should be in the ratios: CH_2 symmetric stretch, 1.379; CH_2 rocking, 1.379; CH_2 asymmetric stretch, 1.349; bending, 1.349; wagging, 1.323; and twisting, 1.414. Where the CH_2 modes are coupled to other vibrations, as in polyethylene, these numbers will not be obtained exactly, but their relative values should be in the same order.

Frequencies observed with isotopic species also differ slightly from calculated values because of a small but appreciable change in bond length. For example, in HCl^{35}, $r_0 = 1.2837$ Å, but for DCl^{35}, $r_0 = 1.2813$ Å. The difference of 0.002 Å is significant. In general, substitution of a heavier isotope results in shorter bond lengths. This apparent discrepancy arises because the observed bond length is an average one taken over the total vibration; heavier isotopes make smaller excursions and thus show a smaller average bond distance. The equilibrium bond length r_e, calculated for a nonvibrating molecule, is strictly independent of mass.

Geometry

The symmetry and geometry of the molecule also have a marked effect on frequency and intensity of its vibrational absorptions. We recall that a vibration is IR active if it shows a change in dipole moment, and the intensity of an absorption band depends on the change in dipole moment that the incident radiation "sees" during the vibration. Clearly, then, absence of a group-frequency absorption at the appropriate place in the spectrum need not indicate absence of the group. The C=C bond furnishes a good example. Its stretching absorption is absent from molecules such as $H_2C{=}CH_2$ and $Cl_2C{=}CCl_2$, and from *trans* isomers

because the vibration occurs about a center of symmetry and no change in dipole moment results (this vibration is strongly Raman active, however). The same considerations apply to methyl acetylene, which shows a —C≡C— absorption, and acetylene, which does not.

It might seem that even *cis* isomers, such as

$$
\begin{array}{ccc}
H & & H \\
\diagdown & & \diagup \\
 & C{=}C & \\
\diagup & & \diagdown \\
Br & & Br
\end{array}
$$

would show only a weak C=C stretch absorption since a plane of symmetry can be drawn through the C=C bond perpendicular to the plane of the molecule. On the contrary, the molecule shows a strong —C≡C— absorption at 1589 cm^{-1}, because the 1589-cm^{-1} band is only about 80% C=C stretch. The remaining 20% involves motions of the CH and CBr bonds, which give a pronounced change in dipole moment [230]. It should be noted that although the rule of symmetry holds accurately in the gaseous state, in liquids or solids the molecules are perturbed by their neighbors and "forbidden" absorptions may be weakly active.

Group frequencies can also be shifted because of mechanical effects in the vibrating system. The C=C stretch vibration in the series of cycloalkenes provides a good example [56, 180]:

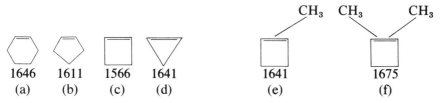

| 1646 | 1611 | 1566 | 1641 | 1641 | 1675 |
| (a) | (b) | (c) | (d) | (e) | (f) |

In cases (a), (b), and (d), the vibrating system must do extra work against the adjoining C—C single bonds; an additional force constant is involved and thus the frequency is raised. In structures (e) and (f), the vibration must work against the =C—CH₃ bond and again the frequency increases. Studies of vibrating mechanical models support these frequency trends [54].

Vibrational Coupling

Group vibrations having approximately the same frequencies and occurring in adjoining portions of the molecule sometimes interact to give a mixed vibration in which both groups take part. Actually, such vibrational coupling is quite common, but this does not prevent the vibrations involved from being good group frequencies. The well-known C=C frequency at approximately 1620 cm^{-1} in substituted ethylenes, for example, is in reality about 60% C=C stretch and 35% CH bending. When the interaction is removed by substitution of deuterium for hydrogen, the C=C frequency drops to its "normal" position at approximately 1520 cm^{-1} [230]. Another example is that of the amide II and III bands, which fall near 1550 cm^{-1} and 1290 cm^{-1} and are quite characteristic of open-chain secondary amides. These bands arise from mixed vibrations involving NH deformation and C—N stretching [93]. The degree of coupling does not vary much from one structure to another and the resulting frequencies are fairly constant.

Conditions that must be fulfilled for vibrations to interact in this manner are: (1) both vibrations must have the same symmetry; (2) both vibrations must have approximately the same frequency; and (3) interactions between groups must be appreciable (this statement implies that the groups are in close proximity and that a mechanism exists for transferring vibrational energy from one group to the other).

A special case of vibrational coupling occurs when an overtone or combination frequency interacts with a fundamental vibration. This type of interaction is called *Fermi resonance*, after E. Fermi, who first proposed such a mechanism to explain the splitting of the 1330-cm^{-1} fundamental of CO_2 [86]. Such interaction results in two bands where one (the fundamental) is expected. The overtone or combination band borrows intensity from the fundamental, and the two bands repel each other so neither is found where expected. The relative intensities and the separation of the bands depend on how closely the two unperturbed frequencies fall. These positions may be calculated approximately using the expression

$$\nu_0 = \tfrac{1}{2}(\nu_a + \nu_b) \pm \tfrac{1}{2}(\nu_a - \nu_b)\frac{I_a - I_b}{I_a + I_b} \tag{5.26}$$

where the two values of ν_0 are the unperturbed frequencies; ν_a and ν_b are the observed frequencies; and I_a and I_b are their intensities [165, 246]. Another example of Fermi resonance is the interaction of the CH stretch in aldehydes at around 2800 cm^{-1}, with the first overtone of the 1400-cm^{-1} in-plane CH bending to give a characteristic doublet near 2700–2900 cm^{-1}.

Fermi resonance is not always easy to diagnose, but it should be suspected whenever a normally single band is split. Sometimes the behavior of the band pair as a function of physical state and solvent will confirm the presence of Fermi-type interactions [30, 33, 262]. In methyl compounds, Fermi resonance between the CH deformation overtone and CH stretch modes has been circumvented by partial deuteration of the group to give CHD_2. Other means of assessing Fermi resonance have also been reviewed [190].

Bond Order

We noted in Gordy's rule (equation (5.24)) that the force constant of a bond varies directly with the bond order. The force constant for C—C, for example, is about 4.6×10^{-5} dynes/cm; for C=C, about 10×10^{-5} dynes/cm; and for C≡C, about 15×10^{-5} dynes/cm.

As we see later, vibrational frequencies can be correlated with bond length, which is a function of bond order. Bond order need not take on integral values; Pauling [208, p. 239) gives an equation for bond distance $D(n')$ relating to bond order as

$$D(n') = D_1 - 0.71 \log n' \tag{5.27}$$

Bond order n' and bond number n are equal for $n' = 1, 2,$ or 3, but for fractional values of n they are not equal. In benzene, for example, $n = \tfrac{2}{3}$ but $n' = \tfrac{5}{3}$, reflecting the extra resonance energy of the molecule.

The stretching frequency of a bond gives a sensitive measure of its length. For C—O and C—N systems [169], bond length correlates with frequency; linearly for the C—N bond, and in a more complex fashion for C—O bonds, over a frequency range of 800–2400 cm^{-1}. Some individual deviations from the plots are no doubt due to vibrational coupling, but in general the correlation is quite good. In a study of X—H stretching frequencies (X═C, N, and O), Bernstein has shown [32] that the X—H bond distance may be derived from vibrational frequencies with about the same accuracy as from high-resolution IR structure studies. Here again, the correlation plot is nonlinear. Trends in bond length with structure have been studied by McKean et al. [192].

Electronic Effects

The distribution of electrons in the molecule (which determines its chemical properties as well as the force constants between atoms) depends on many factors; some are known and some are not. Attempts have been made to classify electronic interactions into inductive, resonance, and field effects for atoms (or groups) by means of inferences drawn from the chemical or physical behavior of molecules containing those groups. Various kinds of reactivity scales were devised that assigned relative values to substituent groups or atoms. These values were useful to predict rates and equilibria for reactions of new molecules [171].

Much effort has been devoted to correlation of IR group frequencies with substituent constants. The purpose of such correlations, of course, is to enable predictions of group frequency shifts from a knowledge of the substituents; such information can be invaluable in establishing or disproving a postulated structure.

The inductive effect. The ability of an atom in a molecule to attract electrons to itself is measured by its electronegativity. Relative values of electronegativity have been derived for all the elements [107], and although the numbers are significant to only one or two figures, the concept has been a useful one. The inductive effect has its origin in the electronegativity difference between atoms in the same molecule or group.

It is helpful to consider electronegativity as a property of the bond rather than of the atom [219] since the value obtained for an atom varies with its chemical state (e.g., 2.3 for carbon in CH_3 to 2.8 for carbon in C≡N). Chemical groups have also been assigned electronegativity values, and group electronegativities have been related to polar substituent constants [68].

A number of multiple-bond stretching frequencies have been correlated with the sum of electronegativities of substituent groups, including phosphoryl ($R_3P{=}O$) stretching [19], carbonyl ($R_2C{=}O$) stretching [147], and others [65]. An electronegativity scale based on A—H stretching force constants has been proposed [271].

For molecules containing σ-bonds, the *inductive effect* is measured as the effect of substituent X on a well-removed reaction center Y (e.g., $X\overrightarrow{CH_2}\overrightarrow{CH_2}\overrightarrow{CH_2}Y$). A quantitative measure of the inductive factor in aliphatic systems, designated σ^*, has been derived by Taft [251].

The inductive effect operates to change the frequency of a vibrating group by changing the bond order (and length). For example, carbonyl groups are influenced as follows:

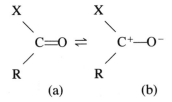

Substituent groups X that reinforce the tendency of the oxygen to attract electrons (oxygen is more electronegative than carbon), i.e., *donate* electrons as in (b), decrease the bond order, and *lower* the carbonyl frequency. Substituents that *attract* electrons, and thus favor (a), *raise* the carbonyl frequency. Acid chlorides, for example, absorb near 1800 cm^{-1} and ketones, near 1715 cm^{-1}.

It does not necessarily follow, however, that the entire shift is due to changes in the $C{=}O$ *force constant*. In almost all studies on the correlation of carbonyl frequencies and intensities, the assumptions have been made that: (1) vibrational coupling of the $C{=}O$ stretching mode with other molecular motions is negligible; and (2) the observed IR frequency of the carbonyl vibration is a direct measure of the sum of the various electrical effects acting on that bond. Overend and Scherer [205], however, have shown in a striking manner that such assumptions are not necessarily true and propose that it is the unperturbed force constant of the bond that reflects its true strength. The carbonyl halides, for example, whose observed frequencies lie in a range covering 100 cm^{-1}, have almost identical force constants (Table 5.1). The value ν^* is a calculated frequency corrected for interactions.

The reason for the apparently negligible difference in electronic effects between F and Br in this dramatic example is not clear, and it should not be inferred from these data on the carbonyl halides that such forces are without influence on the carbonyl frequency. Rather, one important conclusion resulting from similar calculations on other types of molecules is

TABLE 5.1 Force Constants for Carbonyl Halides

Molecule	ν (observed) (in cm^{-1})	ν^*	K$C{=}O$ (in mdynes/Å)
COF$_2$	1928	1861 ± 9	12.61
COBrF	1874	1864 ± 9	12.64
COClF	1868	1848 ± 9	12.43
COBr$_2$	1828	1874 ± 9	12.79
COCl$_2$	1827	1856 ± 9	12.54

that there remains a considerable variation in ν^*_{CO} (and in K_{CO}), which indicates strong dependence on electronic effects. Furthermore, the shift is in the same direction as predicted, but the displacement due to polar effects is not as great as had been supposed.

A good deal of effort has been devoted to studies of inductive effects of substituents on group frequencies. It is important to note that the inductive effect predominates only if the substituent is an appreciable distance away from the vibrating group (e.g., at least one carbon atom removed) [21].

Whereas inductive effects were postulated to transmit through polarization of the chemical bonds [31], a contrary view has been taken by Dewar [73], who proposed that the entire result was due to electrostatic interaction across space (i.e., a field effect). Although some experimental evidence supports this view [245, 257], conventional through-bond interaction is still thought to be a significant, although perhaps not dominant, mechanism for transmission of polar effects.

Mesomerism, resonance, and conjugation. "Mesomerism" and "resonance" both refer to the quantum-mechanical phenomenon whereby a molecular structure is stabilized by contributions from several hypothetical stationary states in such a way that the total energy of the system is minimized. The term "mesomerism" was suggested by Ingold in 1933 to describe the conjugative displacement of electrons in a normal molecular state [134] (e.g., R$_2$N—C=C—C=O). Factors necessary for the occurrence of resonance of this type are: (1) a favorable energy situation for the resonance forms, (2) availability of electrons to form multiple bonds (e.g., lone-pair electrons from oxygen or nitrogen), and (3) coplanarity of the atoms involved in the resonance. Thus we would not expect any appreciable contribution from the form Cl$^+$=C—CO$^-$ in acetyl chloride, since the energy difference between this and the normal form can be calculated to be about 110 kcal (confirming our intuitive conclusion based on relative electron-attracting power).

"Conjugation" refers to the presence of alternate multiple and single bonds. Resonance or rehybridization occurs, with the result that the multiple bonds transfer some of their π-electron character to the intervening single bond. The vibrational frequency of the single bond then increases and that of the multiple bond decreases. The observed lowering in the case of conjugated C=C is about 20–40 cm^{-1}. A splitting of the absorption also results, corresponding to the in-phase and out-of-phase stretching of the C=C bonds. The intensity of the absorption also increases. Such observations need not necessarily imply changes in bond order, in view of the appreciable vibrational coupling that occurs between C=C stretch and CH bending frequencies.

The effect of resonance on the carbonyl band is to lower its frequency by 30–40 cm^{-1} (cf. acetone, 1715; and acetophenone, 1685 cm^{-1}). It is postulated that the electron density in the carbonyl π-bond is reduced by interaction with an adjacent π-bond:

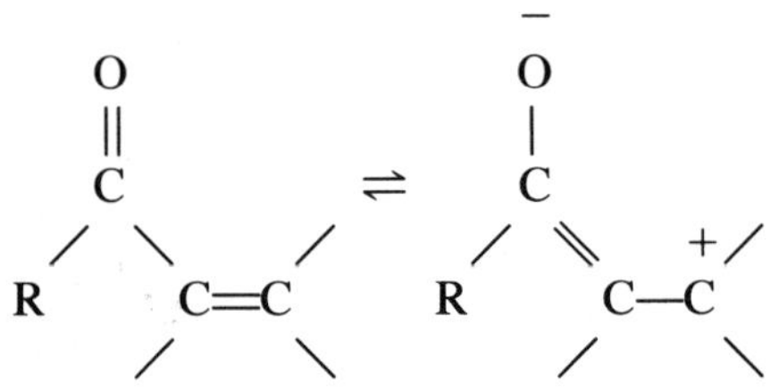

The inductive impact of the atoms involved in such resonance is undoubtedly changed also. Since the conjugative effect may be opposed or reinforced by changes in the inductive effect, prediction of its magnitude is not always easy. Further discussion of the effects observed in molecules with conjugated bonds is given elsewhere [22, 24].

The Hammett equation was developed to correlate structure with reaction rate and equilibrium data for *meta*- and *para*-substituted derivatives of benzene. It provides a measure of the electronic effect of a substituent on a reaction center in aromatic systems. The equation relating the substituent effect σ with the reaction rate (or equilibrium) is

$$\log \frac{k}{k_0} = \sigma\rho \qquad (5.28)$$

where k_0 applies to the reference state (the unsubstituted benzene derivative) and ρ is a proportionality constant that measures the susceptibility of a reaction to polar substituents. Typical correlations of IR data with

Hammett σ values have included the frequencies of the 3, 2, or 1 adjacent-ring hydrogen atoms [20], the carbonyl frequencies of acetophenones and benzophenones [96, 144], the absorption intensities of the amine NH stretching bands [160], and the intensities of the C≡N absorptions in aromatic nitriles [40].

Substituent constants and IR frequencies. A large number of other sub-substituent constants (e.g., σ_p^-, σ_p^+, σ_G, σ_m^+, σ_I, σ_R^0, to name a few) have been proposed to correlate reaction rates, equilibrium constants, and other properties of various families of benzene derivatives. Spectroscopists have spent much time in developing correlations with such parameters, with the parameter usually chosen to give a good fit with the data rather than for any logical reason related to chemical structure.

It has been pointed out by Swain and Lupton [248] that the proliferation of substituent constants had reached the point of absurdity, and they suggested that *any* substituent constant can be represented by a linear combination of resonance and field effects (the latter is defined to include inductive effects). That is,

$$\sigma = fF + rR \qquad (5.29)$$

where F and R are field and resonance constants characteristic of the substituent (e.g., H, OMe, Br, NH_2), and f and r are weighting factors that are independent of the substituent but different for each set of substituent constants (σ_m, σ_p, σ^+, σ^1). While this concept has proven valuable, the use of only a single resonance parameter has been questioned [42].

More recently, the tendency has been to define polar (or inductive) effects as the *result* of effects arising from unconjugated, sterically remote substituents on an equilibrium or rate process, regardless of whether the effect is transmitted through bonds or through space. The end result is measured by a σ_I scale, which is derived from a statistical analysis of reactivity, equilibrium, and physical property data from many sources [257]. The values of σ_I form a series that is similar to but not identical with that described by Taft's σ^*. Inductive scales for *alkyl* substituents are almost identical regardless of the model used to obtain them [173].

Scales for resonance effects have been similarly derived, and four separate scales are given, each one appropriate to a different situation depending on whether the benzene ring is unperturbed (σ_R^0), electron-poor (σ_R^+), electron-rich (σ_R^-), or in resonance with a carboxylic acid group (σ_R) [257]. Constants for some common substituents are given in Table 5.2.

TABLE 5.2   Substituent Constants[a]

	σ_I	σ_R^-	σ_R°	σ_R	σ_R^+
NMe$_2$	0.06	−0.34	−0.52	−0.83	−1.75
OMe	0.27	−0.45	−0.45	−0.61	−1.02
F	0.50	−0.45	−0.34	−0.45	−0.57
Cl	0.46	−0.23	−0.23	−0.23	−0.36
Me	−0.04	−0.11	−0.11	−0.11	−0.25
Ph	0.10	0.04	−0.11	−0.11	−0.30
COMe	0.28	0.47	0.16	0.16	0.16
CO$_2$R	0.30	0.34	0.14	0.14	0.14
NO$_2$	0.65	0.46	0.15	0.15	0.15
H	0.00	0.00	0.00	0.00	0.00
Et	−0.05				
t-Bu	−0.07				
OAc	0.39				

[a] Data from Brownlee and Topsom [42].

The overall effect on the molecule is then given by the dual-parameter equation (where the frequency shift is $\nu - \nu_0$):

$$\nu - \nu_0 = \rho_I\sigma_I + \rho_R\sigma_R^- \tag{5.30}$$

The relationship of IR frequency shifts to reactivity parameters (inductive and resonance) has been considered for many literature correlations—by Brownlee and Topsom [42] for aliphatic compounds and by Brownlee, Di Stefano, etal. [41] for aromatic molecules. In most cases, the use of the dual-parameter approach gave better correlations. These authors stress the importance of testing a wide range of substituents in establishing correlations. The inductive effect has been reviewed [173, 245] and substituent electronic effects discussed [257]. Some examples of correlations found in the literature are now given.

a. The NO$_2$ asymmetric stretching mode in aliphatic nitro compounds [181]. In compounds of the type R(CH$_3$)$_2$CNO$_2$, the frequency of the NO$_2$ asymmetric stretch vibration is given approximately by the equation

$$\nu_{(asym.)} = 23.92\ \sigma^* + 1538.5\ \text{cm}^{-1} \tag{5.31}$$

with a correlation coefficient of 0.981.

b. The G—CH=CH$_2$ twist frequency [217]. A study of 69 substituted vinyl compounds has shown a correlation of the HC=CH$_2$ twist frequencies with the pK_a values of the correspondingly substituted acetic acids (Fig. 5.8). The point for C≡N falls off the curve because of hydrolysis of the C≡N group, so that an accurate value for the pK_a of cyanoacetic acid is not obtained. Points for methyl, ethyl, and isopropyl groups may deviate from the correlation line because of vibrational coupling. Although the reason for the variation of the CH=CH₂ twist frequency with inductive power of substituents is not clear, the relationship is a useful one.

c. The C=O stretching frequency of methyl benzoates [69]. The carbonyl stretching frequencies in a series of *meta*-substituted methyl benzoates in dilute CCl₄ solution, when correlated with the single Hammett substituent parameter σ_m, gave the result shown in Fig. 5.9. Use of the dual substituent parameter treatment gave the equation

$$\nu_m - \nu_0 = 12.79\,\sigma_I + 5.63\,\sigma_R \qquad (5.32)$$

which fits the points with about half the standard deviation calculated from an σ_m correlation. Substituents were chosen to include a wide range of properties and included NMe₂, OMe, F, Cl, Br, Me, H, CO₂Me, and NO₂.

The intensities show analagous trends; the plot of a single-parameter

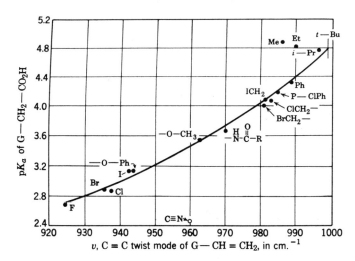

Fig. 5.8 Twist frequency of C=C in G—CH=CH₂ compounds plotted against pK_a values of corresponding acids GCH₂COOH. Reproduced, with permission, from *Spectrochimia Acta* [217].

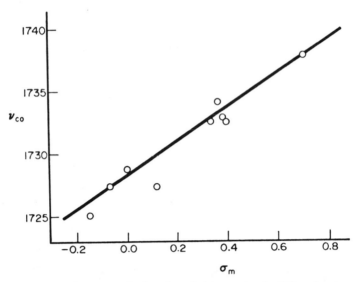

Fig. 5.9 Plot of ν_{CO} for *meta*-substituted methyl benzoates in CCl_4 solvent against σ_m values. Reproduced, with permission, from *Spectrochimica Acta* [69].

fit with σ_m is shown in Fig. 5.10. The dual-parameter equation

$$A_m^{1/2} - A_0^{1/2} = -8.80\,\sigma_I - 0.50\,\sigma_R \qquad (5.33)$$

gave somewhat less scatter.

Correlations between IR frequencies and substituent constants have been discussed at length by Topsom and co-workers [41, 42, 152].

Summary. Our discussion of electronic effects can be summarized as follows:

Inductive effects: (1) follow in order of σ_I or Taft σ^* substituent constants, with the largest constants having the greatest effect; (2) change stretching frequencies in proportion to the magnitude of the constant; and (3) correlate well with frequency shifts in cases in which the substituent is at least one or two atoms removed from the vibrating group.

Conjugation of multiple bonds: (1) lowers the frequency and (2) raises the intensity of the multiple-bond stretching vibration.

Resonance: (1) depends on the availability of electrons for multiple-bond formation; (2) is most likely to occur when the atoms involved lie in a plane; (3) inevitably changes the inductive effect of the atoms involved; and (4) has a large effect on frequency, with the

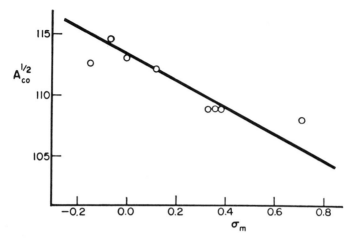

Fig. 5.10 Plot of square root of absorption intensities for *meta*-substituted methyl benzoates in CCl_4 against σ_m values; A values are in 1-mole^{-1} cm^{-2}. Reproduced, with permission, from *Spectrochimica Acta* [69].

magnitude and direction of the shift as a function of the bond order and the secondary inductive effect.

It is evident at this point that correlations of frequency or intensity shifts with inductive or resonance parameters, although useful from an empirical standpoint, should not be used to draw far-reaching conclusions about molecular structure. Not only are inductive, resonance, and mass effects entangled in a complicated and often obscure fashion, but the use of frequencies rather than force constants can lead to erroneous conclusions about how these effects operate.

Meanwhile, the analytical spectroscopist cannot be expected to work out normal coordinates for every molecular species he encounters. In analytical work the empirical approach is doubtless best for the present, and additional correlations of frequencies or intensities with other physical properties should prove extremely useful.

Association Effects

In a broad sense, molecules in any condensed phase are associated, and the interactions of solute with solvent or of the pure material with itself, will produce changes in the vibrational spectrum. This type of interaction is discussed later. The discussion given here is confined to discrete interactions that occur in addition to van der Waals attraction.

One type of self-association is that found in compounds that form electron-deficient bonds, such as $Al_2(CH_3)_6$ and $Be_x(CH_3)_{2x}$. Alkyl lithium compounds are particularly prone to form polymers of this type [264], which persist in the vapor state [265]. Some metal alkoxides also form aggregates [15]. In all these cases the IR spectra of the polymers show some rather striking differences from the pattern expected for the monomers.

Lewis acids form complexes with carbonyl compounds. The shift in the C=O frequency of acetone complexed with BF_3 is 70 cm^{-1}; for acetophenone it is 107 cm^{-1} [59, 247]. These shifts seem to correlate with the basicity of the C=O group, as might be expected. Ethers form complexes with HCl and HF that persist in the vapor [8].

Hydrogen bonding is by far the most common associative effect. It involves a special type of interaction between a proton-donor molecule XH and a proton-acceptor molecule. The usual donors and acceptors are electronegative atoms such as —OH···O—, —OH···N—, —NH···O—, and NH···N— groups (or in some cases, π-electrons). Intermolecular interactions of CH (as in $HCCl_3$ solutions) are also sometimes described as hydrogen bonding.

The effect of hydrogen bonding on the IR spectrum is to shift and broaden the XH stretching absorption, as shown in Fig. 5.11. The integrated intensity also increases markedly. A completely satisfactory explanation for these phenomena has yet to be proposed, but a number of useful observations have been made, which are reviewed briefly.

Criteria for the formation of hydrogen bonds have been listed by Cannon [47]: (1) the X—H bond must be partially ionic in character (or such that ionic character can be induced by polarization); (2) the acceptor atom must have lone pair electrons in an asymmetric orbital; and (3) for maximum interaction, the X—H bond and the lone-pair orbital axis must be collinear. (It is *not* implied that the *bonds* must lie on a straight line, viz. —O—H···O—.) It might be added that steric hindrance of bulky groups surrounding a proton donor site will inhibit hydrogen bond formation.

The heat of formation of the ordinary intermolecular hydrogen bond is usually 3–10 kcal/mole. The greatly broadened absorption often consists of several overlapping bands corresponding to equilibrium concentrations of dimer, trimer, and other polymeric species (Fig. 5.12). The relative amounts of these different species depends on the solute concentration, the solvent, and the temperature. Not until dilutions on the order of 10^{-2} to 10^{-3} molar are approached do concentrations of polymeric species become negligible. Because of this strong dependence of intensity on

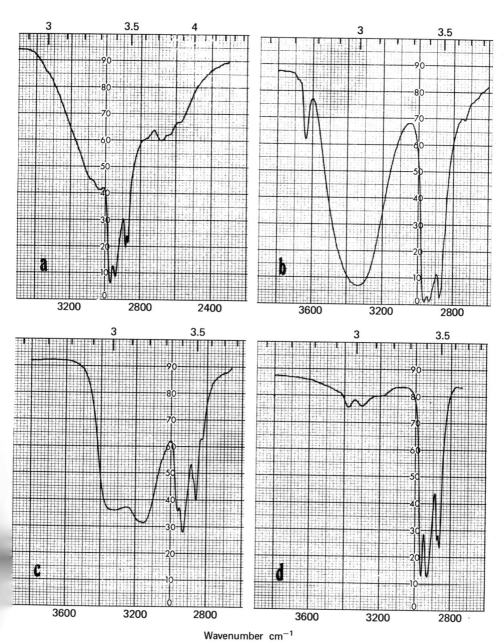

Wavenumber cm⁻¹

Fig. 5.11 Hydrogen bonding absorption in some typical associative groups: (*a*), 2-ethyl-hexoic acid; (*b*), *n*-propyl alcohol; (*c*), acetamide; (*d*), *n*-hexylamine. Spectra (*a*) and (*b*) are 10% solutions in CCl₄, (*c*) is a mull, and (*d*) is a liquid film.

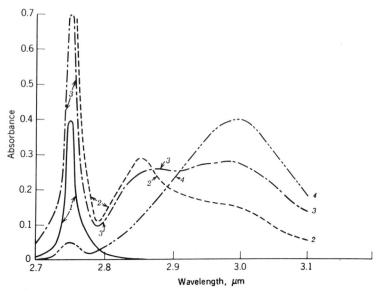

Fig. 5.12 Distribution of H-bonded *n*-butanol species in CCl$_4$: curve 1, 0.005 *M* in 20-mm cell; curve 2, 0.075 *M* in 3-mm cell; curve 3, 0.152 *M* in 1-mm cell; and curve 4, 1.7 *M* in 0.025-mm cell. Reproduced, with permission, from *Spectrochimica Acta* [164].

external factors, OH and NH absorptions arc not suitable for quantitative analyses except in special instances.

Heats of formation for various aggregate species can be calculated by studying their concentration dependence on temperature. Such studies have been carried out for some aliphatic [174] and aromatic alcohols [184], and typical results are shown in Table 5.3.

TABLE 5.3   Hydrogen-bond Energy of Alcohols in CCl$_4$ by IR

Alcohol	Species	Energy (in kcal/mole)
Methanol	Dimer	9.2 ± 2.5
	Polymer	4.7 ± ?
Ethanol	Dimer	7.2 ± 1.6
t-Butanol	Dimer	4.8 ± 1.1
	Polymer	5.3 ± 0.5
Phenol	Dimer	5.12 ± 0.10
p-Cresol	Dimer	6.09 ± 0.18
p-Cl-phenol	Dimer	3.78 ± 0.20

Some molecules form hydrogen-bonded dimers and polymers even in the vapor. Gaseous HF, for example, is associated; possibly as $(HF)_6$. Acetic and butyric acids form dimers with enthalpies of 8.5 ± 0.5 kcal/ mole [220].

The absorption frequency shift from the unassociated to the associated species has been correlated with the X—Y distance in the X—H$\cdots$Y unit [26, 27, 179, 232]. The relationship is such that the shifts are greater the closer is the approach of X to Y.

The frequency shift of a given proton donor to a variety of bases (proton acceptors) can be related to the heat of formation of the hydrogen bonds [10, 108]. By way of illustration, enthalpies of some phenol-acceptor systems are: benzene, 1.6 kcal; ethyl acetate, 4.5 kcal; diethyl ether, 5.4 kcal, and ethyl methyl ketone, 5.3 kcal [218]. The band widths and integrated intensities also correlate with enthalpy, at least for certain compounds [18].

It might be anticipated that the proton involved in hydrogen bond formation moves in a potential well with a double minimum, corresponding to the positions X—H$\cdots$Y and X$\cdots$H—Y, with a maximum at the point X$\cdots$H$\cdots$Y. Barrow [16] has established tentative criteria for the existence of the double minimum. The effect on the spectrum is to split certain energy levels, with the result in some cases that the fundamental or overtone bands are doubled.

In addition to the *inter*molecular hydrogen bonds discussed previously, some molecules have structures that permit the formation of *intra*molecular hydrogen bonds to adjacent groups on the molecule [164]. (Certain *ortho*-substituted aromatics such as orthochlorophenol,

fall into this class.) Since association is sterically favored, the heat of formation may be lower (1–4 kcal, estimated). The shifts and band widths are smaller and are not necessarily related to the heat of formation of the bond. Furthermore, the protons may associate with weaker bases, such as π-electrons in the allyl group of 2-allyl phenol [14], where the observed shift on association is 63 cm^{-1} (Fig. 5.13). The association of the intramolecular bonded protons persists even at extreme dilutions.

Despite the variability of the associated hydrogen absorption, the band is of some use in qualitative analysis. The extremely broad absorption with numerous minima shown by fatty acids (Fig. 5.11) is certainly char-

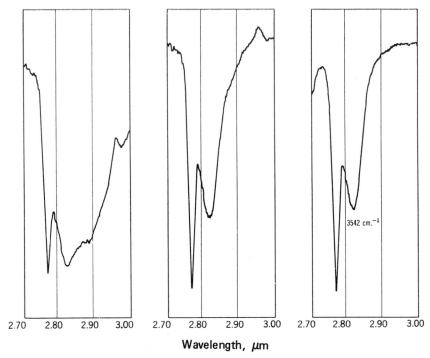

3542 cm.$^{-1}$

2.70   2.80   2.90   3.00   2.70   2.80   2.90   3.00   2.70   2.80   2.90   3.00

Wavelength, μm

Fig. 5.13 Spectra of the OH stretching vibration of 2-allylphenol at 0.7 M, 0.07 M, and 0.007 M. Reproduced, with permission, from *Journal of the American Chemical Society* [14].

acteristic. Self-associated amines and alcohols can sometimes, but not always, be distinguished; in the unassociated state differentiation is easier (Appendix 2). Water may interfere, if present, but it can often be detected by its deformation band near 1620 cm^{-1} (Fig. 5.14).

Other vibrations also may be affected by the hydrogen bond. If, for example, the proton acceptor is a carbonyl group, its vibrational frequency will be lowered by amounts up to 50 cm^{-1}, and in some cases the association may be studied more conveniently by its effect on the base.

Internal vibrations of the donor molecule also change. In an alcohol, for example, we expect C—O stretch, OH deformation, and COH torsion modes to be more or less affected by hydrogen bond formation. These modes are often mixed and thus may be difficult to characterize. The vibrations of associated donor molecules have been discussed in some detail by Pimental and McClellan [211], who arrive at the following generalizations: (1) X—H in-plane bending mode falls in the region 1000–

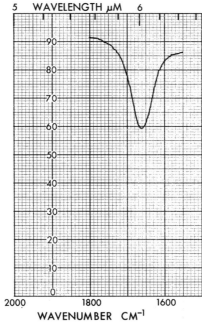

Fig. 5.14   Deformation band of water in ethanol.

1700 cm^{-1} and shifts upward with hydrogen bond formation; (2) the X—H out-of-plane torsion lies below 800 cm^{-1} in associated liquids and solids; (3) the X—H$\cdots$Y stretching and bending modes are low in frequency (on the order of $\leq$200 cm^{-1}); and (4) the C—O stretch in alcohols falls near 1000–1070 cm^{-1} and is lowered by hydrogen bond formation.

For further discussion of the many experimental and theoretical studies on hydrogen bonding, the reader should consult the literature [24, 111, 137, 159, 197, 210, 228, 238].

External Influences

The term *external influences* is somewhat ambiguous and the division from internal effects is rather arbitrary, but in this case external effects are taken to mean these influences that are, to some extent at least, controllable by the investigator. This category includes physical state (gas, liquid, solid, solution), solvent, concentration, and temperature. Each of these variables has an important influence on the position of a group frequency and, in order to achieve maximum use of group-frequency correlations, one should be well aware of the possibilities for external perturbations.

Physical State

Nowhere is the effect of molecular environment on the IR-absorption pattern of a chemical species more strikingly demonstrated than on passing from gas or vapor to a condensed state (Fig. 5.15). In the gas, the individual molecules are free to vibrate and rotate with little perturbation from other molecules. The resulting spectrum, as we have seen (pp. 129–

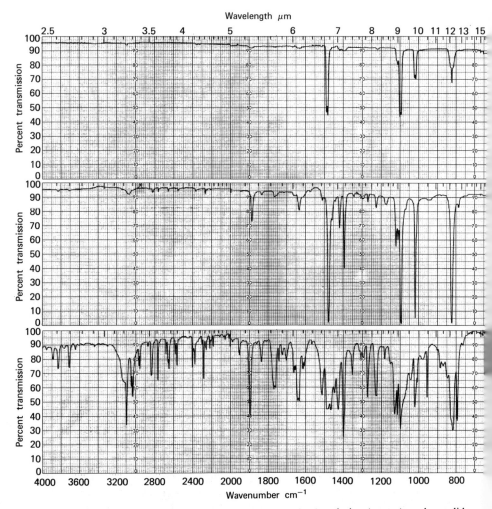

Fig. 5.15 Spectra of *p*-dichlorobenzene, run as vapor (top), solution (center), and as solid resolidified melt (bottom).

131), is a series of absorptions, each consisting of many narrow bands corresponding to individual rotational transitions superimposed on the vibrational transition, each covering a broad range of wavelengths. In liquids or solutions, the individual molecules are confined to a molecular cage where they are continually buffeted by other particles so that they can no longer execute quantized rotational motion. As a result, the fine structure of the vibrational band disappears and the absorption takes on a shape somewhat similar to a probability function. Forces responsible for strong local perturbations include dispersion forces, dielectric effects, dipole–dipole and dipole–induced-dipole interactions, and specific associative phenomena such as hydrogen bonding.

The magnitude of the shift on passing from gas to liquid is rather variable. It may range as high as 100 cm^{-1}, but ordinarily it is less than 25 cm^{-1}. The C=O stretch in ethyl acetate vapor, for example, falls at 1763 cm^{-1}, and in CCl_4 solution at 1741 cm^{-1}. The C=O stretch in acetone falls 18.5 cm^{-1} on going from vapor to CCl_4 solution. The acetylenic C—H stretch is quite sensitive to its physical state, possibly because of association in solutions through a hydrogen-bond bridge. Phenyl acetylene shows absorption at 3340 cm^{-1} in the vapor and 3316 cm^{-1} in CCl_4 solution. The problem has been treated theoretically for symmetrical pentatomic molecules by Heicklen [116], who finds that bond lengths increase on going from vapor to liquid.

In the 500–4000 cm^{-1} region, frequencies almost always are lowered on going to a condensed phase. It has been found, however, for some low frequencies (in the 200–340 cm^{-1} range), that liquid-phase values are higher than gas-phase ones [84].

In the crystalline state, force fields acting on individual molecules are coherent, with the result that a further change in the spectrum occurs (Fig. 5.15). The vibrational absorptions usually become sharper and often split, and new bands may appear. This happens because each unit cell acts as a vibrational entity, but a unit cell usually contains more than one molecule. Thus the possibility exists for in-phase and out-of-phase motions, which may have different frequencies. Libration (partial rotation) of the individual molecules can also occur, as well as lattice vibrations of low frequency, and these energy states are superimposed on the vibrational absorption. Also, the strong local force fields may change the frequency of a vibrational band or may render a normally inactive mode IR active.

Molecules such as fatty acids may take a preferred configuration in the crystalline state, giving rise to characteristic bands that are useful for determining chain length [194]. Rotational isomerism in substances such as 1,2-dichloroethane often disappears as all isomers take on the lowest energy state on solidification, with a resulting simplification of the spectrum [234]. The phenomenon of polymorphism is well known, and different crystalline forms of the same species often show striking differences in their IR spectra. Some liquids supercool to form glasses. The spectrum of a glass is usually not very different from that of the liquid. The spectra of single crystals of a number of substances have been studied and interpreted in terms of the lattice structure. Selection rules have been developed to predict the activity of the vibrations [85]. The effects of phase and pressure changes on vibrational spectra have been discussed by Davis [67].

The important implication for the spectroscopist is that aside from deliberate exploitation of solid-state effects, he had best eliminate as many variables as possible and, wherever feasible, run solid samples as solutions.

Solvent

The fact that group frequencies change in both position and intensity on going from one solvent to another is perhaps not as widely appreciated as it should be. The reasons for these shifts are just beginning to be understood, and in this section we review the principal influences acting on dissolved species.

Historically, the first serious attempt at explaining the drop in frequency usually observed in liquids was made on rather meager experimental data by Kirkwood [156] and later by Bauer and Magat [17]; this was subsequently known as the "KBM" theory. They treated the absorbing species as a simple diatomic vibrator in a cavity of solvent with dielectric constant ϵ, and attributed the shift to instantaneous induced polarization of the solvent medium by the vibrating dipole. Their expression, known as the KBM relationship, relates the frequency shift to the dielectric constant of the medium. As we see later, this approach is a gross oversimplification of the actual situation even for nonpolar solvents. More refined expressions [44, 221] have met with only partial success.

Subsequent tests of the KBM theory showed it to be inadequate, and studies have been made to determine the effect of different solvents on a particular frequency, such as the $C=O$ stretch vibration of acetone [256]. This vibration, for example, varies from 1707 cm^{-1} in CH_2I_2 to 1722.5 cm^{-1} in n-hexane; its integrated intensity changes by a factor of 1.6 between n-hexane and pentachloroethane; and its width ranges from 13 cm^{-1} in CH_2Cl_2 to 19.5 cm^{-1} in CS_2. The $C\equiv N$ stretch in p-methylbenzonitrile, however, shows only a small frequency shift in different solvents, but its intensity changes by a factor of 2.3 and its width by a factor of 1.6 [39]. Examples of this type illustrate the variations to be expected among solvents, but they are not very helpful in disentangling the various factors that lead to the changes.

It has also been proposed [30] that associations with solvent molecules are largely responsible for the frequency shifts and that dielectric effects have relatively little importance.

The most rewarding approach to the study of solvent shifts has been through high-resolution studies of a single solute in two solvents mixed in varying proportions [25]. If the solvent effect is a simple function of some bulk property (e.g., a dielectric constant), the frequency and intensity should vary smoothly as solvent composition is changed. If, on the other hand, specific interactions with solvent molecules occur, two bands should be in evidence, with their intensity changing as the solvent composition is varied.

A systematic study of this type has been carried out by Kagarise [148] and Whetsel [267]. Using cyclohexanone and acetone as solutes and p-cresol (with which association is expected) and cyclohexane (which should be relatively inert) as the solvent pair, they were able to show not only that specific interactions occurred with the carbonyl group of the ketones (Fig. 5.16), but also that 1:1 and 1:2 complexes were formed, and to calculate their formation constants. The frequency shift of the C=O stretch mode in cyclohexanone was 16 cm^{-1} for the 1:1 complex and an additional 5 cm^{-1} for the 1:2 complex. Changes in band intensities and widths also were noted.

Using a less active solvent pair, chloroform and carbon tetrachloride, these authors found evidence for 1:1 and 1:2 complexes of the ketones with chloroform, in which the chloroform molecules are probably hydrogen bonded to the carbonyl group. Frequency shifts of the carbonyl absorption due to association were about 8 cm^{-1} for cyclohexanone and 4 cm^{-1} for acetone. Additional small displacements on the order of 3–4 cm^{-1} were found and were attributed to nonspecific interactions (including bulk dielectric effects).

One interesting conclusion emerging from a study of the solvent pair cyclohexane–carbon tetrachloride was that a weak but distinct pairing occurs between cyclohexanone (and presumably other polar molecules) and carbon tetrachloride. The total solvent shift was found to be 6.8 cm^{-1}, of which about one-half was attributed to specific interaction.

Whetsel and Kagarise conclude that both specific solvent–solute interactions and nonspecific interactions, such as bulk dielectric effects and dispersion forces, contribute appreciably to the shift of the carbonyl absorption of a single species in different solvents. The relative impor-

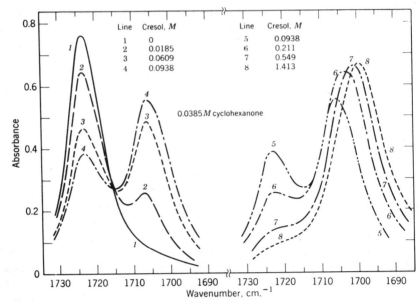

Fig. 5.16 Spectra of the C—O stretch of cyclohexanone in mixtures of *p*-cresol and cyclohexane. Reproduced, with permission, from *Spectrochimica Acta* [267].

tance of the two types of interactions apparently depends on the energy of the associated solvent–solute pair, with strong dipoles giving maximum specific interactions and the largest frequency shifts.

The magnitude of the shift depends also on the particular vibrating group involved. It is easy to see, for example, that association of an X—H group with a solvent acting as a Lewis base depends on the acidity of the proton, and a linear relationship between the N—H stretch frequency in substituted anilines and naphthylamines [43] and their pK_a values has been found. Groups such as carbon–halogen might be expected to show somewhat smaller specific interactions, and, indeed, it has been found that about half the solvent shift for this group results from dipolar interactions, with the remainder from nonspecific forces [113]. Interactions with solvent play a large part in intensity variations of the NH and OH stretch in diphenyl amine and α-naphthol [201] and presumably in other polar groups.

It is clear, then, that frequency changes of a molecule in solvent depend on both the solvent and the group involved [112]. This point is well illustrated by a study of a rather extreme case, *p*-dioxane in water solution [95]. Some of the shifts observed, and their direction, on going from pure *p*-dioxane to dilute water solution are:

$\nu_{11}(B_uCH$ stretch), $+ 19$ cm^{-1}; $\nu_{16}(B_u$ rock), $+ 4$ cm^{-1}

$\nu_{17}(B_u$ ring stretch), $- 5$ cm^{-1}; $\nu_{21}(A_uCH$ stretch), $+ 12$ cm^{-1}

$\nu_{26}(A_u$ ring stretch) $- 1$ cm^{-1}; and $\nu_{27}(A_u$ ring stretch), $- 4$ cm^{-1}

This strikingly varied behavior of different group frequencies in different solvent situations is the basis for an elegant method of group-frequency characterization, which is considered in greater detail later.

It is now clear that no solvent is completely inert; even a solvent as innocuous as CCl$_4$ has been implicated in specific association with solute molecules. On this basis, it might be argued that the use of solvents should be avoided altogether in running spectra. This viewpoint, however, overlooks the fact that strong intermolecular interactions are likely to occur when a liquid acts as its own solvent. Furthermore, since impure or mixed samples are the rule rather than the exception, interactions of the species present are likely to result in much larger frequency and intensity changes than are found with relatively inert solvents. Thus use of the solvent pair CCl$_4$–CS$_2$ for routine work, as recommended earlier (pp. 73–76), is likely to give the most invariant and reproducible group frequencies, intensities, and band shapes.

Concentration

Gases. The IR-absorption spectrum of a gas at low pressure consists of a great many extremely sharp, narrow bands corresponding to transitions between individual vibration–rotation energy levels. Since most of these lines are beyond the resolution capabilities of the usual IR spectrometer, the absorption envelope observed usually consists of an average absorption, with individual transitions observed only for molecules with widely spaced rotational levels, such as HCl. As the total pressure is increased (by addition of a transparent foreign gas), the number of collisions with molecules of the absorbing species increases. The molecule cannot rotate freely during these collisions, and if its rotational energy changes by absorption of radiation, one or both of the energy levels involved are likely to be randomly displaced. Consequently, the average absorption band is widened, the spectrometer is better able to resolve it, and an apparent increase in its intensity results. This phenomenon, known as *pressure broadening*, is of considerable importance in the spectroscopy of gases and vapors.

The effect of adding nitrogen to a constant pressure of methane gas is shown in Fig. 5.17. The absorbance increases drastically during the initial

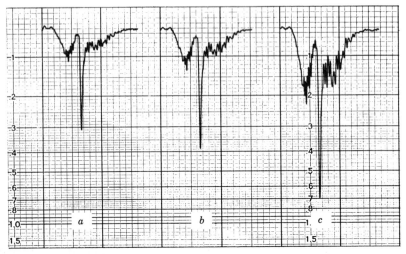

Fig. 5.17   Pressure broadening of 1310-cm⁻¹ band of CH₄: (*a*) 50 mm of CH₄; (*b*) 50 mm of CH₄ plus 40 mm of N₂; (*c*) 50 mm of CH₄ plus 700 mm of N₂.

addition of nitrogen and less markedly at higher pressures. A pressure increase from 0 mm to 700 mm of inert gas may change the intensity of an absorption band by a factor of 5 or more.

The effectiveness of foreign gas molecules in producing pressure broadening depends to some extent on physical size, with larger molecules giving more effect. No rule can be laid down as to order of effectiveness, however. It has been found, for example, that the 1310-cm⁻¹ (7.65-μm) band of methane responds to foreign gases in the order He < A < H₂ = O₂ < CO₂ < C₂H₆ < C₃H₆ [53]. For the NO stretch in N₂O, however, the order is A < O₂ < N₂ < C₂H₆ < CO₂ = H₂. Furthermore, different absorption bands of the same gas may respond differently to pressure broadening. Figure 5.18 shows the behavior of several acetylene bands as a function of the partial pressure of CO₂ [5].

The state of knowledge regarding pressure broadening is not very satisfactory at present. About all that can be said with certainty is that the absorption enhancement depends on the species studied, the specific absorption band observed, and on the gas used to pressurize the systems, as well as on its pressure. From a practical standpoint, there is considerable advantage in running all gases as "solutions" in nitrogen at atmospheric pressure. Such a procedure minimizes intensity variations from pressure broadening and gives better sensitivity for weak absorbers. It also precludes intensity changes resulting from air leaking into the cell.

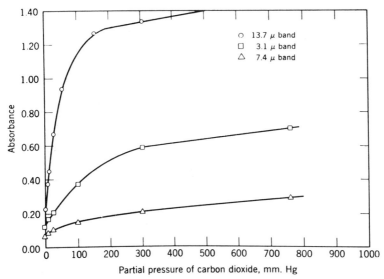

Fig. 5.18 Pressure broadening of several acetylene bands by carbon dioxide. Reproduced, with permission, from *Applied Spectroscopy* [5].

Liquids and solids. In cases where hydrogen bonding or other types of intermolecular association occur, certain absorptions may be quite concentration sensitive (see Fig. 4.1). The reasons for this behavior have been discussed earlier. It is recommended that standardized conditions of solvent, concentration, and path length be adopted for all samples of a given type.

Temperature

The effects of large temperature changes on spectra are well known. At 78 K, for example, at which most materials are solids, absorption bands are sharper, integrated intensities are different, and new bands may appear in the spectrum. At temperatures above ambient, peak-absorption intensities decrease and band widths increase. In gases, changes are seen in band shapes. Even small temperature changes may markedly affect the spectra of associated molecules. Some of the reasons for this temperature dependence are discussed here.

In gases, the distribution of molecules in the various rotational (and vibrational) energy levels follows the Boltzmann distribution, which is a function of temperature. As the temperature increases, the proportion of molecules in the higher-energy states increases and the contours of the vibration–rotation band envelopes change accordingly (Fig. 5.19). More frequent collisions induce broadening of the bands. If rotational isomers are possible, a temperature change may result in a different distribution of isomers, with concurrent variations in the spectra. This situation occurs in 1,3-butadiene [225], in which the *trans* form predominates at room temperature. At somewhat higher temperatures (100°C), the amount of *cis* (or possibly *gauche*) form becomes significant, and changes in several absorption bands reflect this isomer redistribution.

For molecules in the liquid or solution state, temperature changes induce variation in band position, intensity, and width [129, 240]. The CH stretch frequency of chloroform [178], for example, changes position by about 3 cm^{-1} and integrated absorbance by 32% in the interval $-58°C$ to $+60°C$. Self-association of the molecule may affect this frequency, however, and other bands change less drastically in intensity (17–25%). The intensity decreases linearly with increasing temperature, and the effect is attributed to smaller induced dipole interaction as the intermolecular distance increases. Similar effects have been noted with aliphatic compounds [74]. Band widths in toluene, cyclohexane, and acetonitrile were found to change by factors of 1.5–4 over a temperature range $-60°C$ to $+60°C$ [223]. We have already noted the strong temperature dependence of hydrogen-bonded species. Similarly, the distribution of rotational isomers is often affected by temperature, and this dependence provides a convenient method for studying the energy differences between them [234].

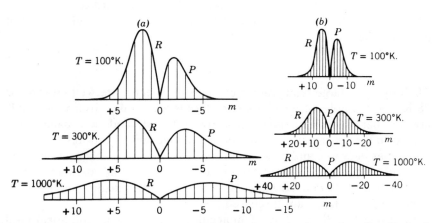

Fig. 5.19 Dependence on temperature of intensity distribution in rotation–vibration bands for: (*a*) HCl, with rotational constant B = 10.44 cm^{-1}; (*b*) molecule with B = 2 cm^{-1}. Reproduced, with permission of Van Nostrand–Reinhold, from G. Herzberg, *Spectra of Diatomic Molecules*, Princeton, N. J., 1950.

Higuchi, Tanaka, et al. [120] have explained the variation in bandwidth as arising from Brownian rotational motion of the solute molecules. Their plots of band widths versus temperature or solvent viscosity are consistent with the idea that more frequent solvent-solute collisions increase band widths, and they have calculated the potential barrier for molecular reorientation for several solutes. This potential barrier, thought to be principally due to specific interaction of the molecule with its nearest neighbors, is on the order of 1.5-3 kcal/mole. Theoretical aspects of reorientational relaxation have reported by Gordon [105] and Shimizu [236]. Experiments with polar solvents [121] have indicated that other less obvious factors may also contribute significantly to IR band widths in solution.

The temperature dependence of the absorption spectra of crystalline materials has not been widely investigated, but as long as changes in crystalline form do not take place, temperature seems to have a relatively small effect on fundamental vibrations. Combination bands between lattice and fundamental modes do show temperature dependence in accord with the Boltzmann factor.

INTERPRETATION OF SPECTRA

The interpretation of IR spectra provides a unique challenge to the chemist. Although the more obvious spectral features can be immediately related to the sample composition, the spectum contains such an immense amount of information that only a fraction of it is utilized even by skilled practitioners. Further, spectra are not "well-behaved," in that recorded band intensities are not proportional to concentrations of the group from which the band originates, and shifts in group frequencies are affected by influences only partly understood. It is nevertheless gratifying that even the newest beginner can, by matching an unknown spectrum with a known one, unequivocally establish the identity of an unknown material. Further, as one's proficiency increases, one can utilize the technique for the solution of structural problems difficult or impossible to solve by other means. In short, IR spectroscopy offers something to everyone.

Although identification of unknown materials and determination of molecular structure are discussed separately, in practice the former problem sometimes resolves itself into the latter. By unknown materials, we mean the usual type of samples encountered in the spectroscopy service laboratory. They run the gamut from distillation cuts to commercial products to odd bits of contaminants of dubious character and uncertain origin. It is safest to assume that all samples of this type are mixtures, or at best, impure.

Structure determination, on the other hand, is carried out on samples that have been carefully purified (it is virtually impossible to deduce structural information on an unknown mixture per se). It is also essential to obtain spectral data that are as complete as possible, including IR absorption spectra over a wide wavelength range, along with Raman shifts and depolarization data. Assistance also may be had in some cases from studies of solvent shifts, vapor-band contours, and other techniques discussed later.

Identification of Unknowns

No cut-and-dried procedure can be given for characterization of unknown materials. One can only suggest approaches that, judging from past experience, have some probability of success. It should also be emphasized that not all unknowns will be identified. In many cases, tentative identification of the major functional groups is all that can be accomplished.

Whereas the chances of success increase with the amount of effort devoted to the solution of a problem, the value of the result must always be weighed against the effort required to achieve that result. There comes a point in every problem at which it is no longer profitable to expend further time and effort. Often the characterization of the major groups present will satisfactorily resolve the problem. Indeed, assurance that certain structures or groups are *absent* often is all that is required. In such cases, complete identification is obviously wasteful. The point is this: it is important to understand the nature of the problem and the proposed use of the solution *before* starting work on the problem.

If we assume that IR is the preferred approach (as it usually is), at least for screening the unknown material, considerable information can be accumulated before the spectrum is run. The physical appearance and properties of the sample are of obvious importance. A colorless crystalline solid, for example, is more likely to be successfully characterized than a colored resinous or tarry material. Viscosity (of fluids), solubility characteristics, approximate melting point, and microscopic examination are examples of tests that may give valuable information. Behavior of a small specimen when inserted into a flame will usually indicate whether the material is organic or inorganic and, if the former, whether aromatic rings are present. More refined flame tests can be used to indicate organometallic compounds [243]. On liquids or volatile solids, a GC analysis will give an indication of the purity of the material. It is dangerous to assume, however, that a single peak on the chromatogram is indicative of a pure sample, because overlap of peaks and nonvolatile residues occur more frequently than many chemists suspect.

The use of sample history is an important part of every identification. A knowledge of all the reactants and reaction conditions, or of the treatment of the material prior to its receipt, often will turn a difficult problem into an easy one. When running samples prepared by other research chemists, the spectroscopist should keep in mind the possibility of accidental contamination by stopcock greases, solvents, extracts from rubber or plastic tubing, and the like. (A list of possible spurious absorption peaks is given in Table 5.4.)

It is wise at this point to determine whether the sample is reasonably pure or has two or more major components. In the latter case, interpreting the spectrum from group frequencies as described in the paragraphs that follow is somewhat more difficult.

The unknown may be a bona fide product of the major reactants, or it may result from an impurity in the starting materials. Infrared examination of the starting materials is well worth the effort and should be a part of every serious attempt to study a new reaction.

After the sample has been run by a suitable technique, as discussed in Chapter 4, its spectrum is critically examined. If the sample was insoluble in organic solvents, an inorganic material should be suspected. Surprisingly, it is not always easy to recognize inorganic materials from their spectra, especially if they have been run by the mineral-oil mull method, which precludes observation of the CH regions. The absorption bands of inorganics are usually broad, but not always.

The hydrogen stretching vibrations are very revealing of the type of compound(s) present. Samples run at standard dilution in solvent as previously recommended disclose by the intensity as well as the position of the CH absorptions whether the molecule is aliphatic, aromatic, or both. Further, the spectrum gives a measure of the amounts of each type of structure.

The spectrum is now considered in the way in which the spectroscopist examines it (i.e., by wavelength regions), and some of the important features of each region are illustrated and discussed. This discussion is not meant to be complete, but rather to point out some typical features of the IR pattern and to illustrate the reasoning the spectroscopist uses when approaching an unknown spectrum.

1. The NH–OH region: 2500–3650 cm^{-1} (2.75–4.0 μm). Unless strong intramolecular hydrogen bonding is present, unassociated OH and NH stretching absorptions usually can be seen when the sample is diluted in an inert solvent. With moderately good dispersion and frequency accuracy, one can distinguish —OH from —NH or —NH$_2$ and can often even discriminate between primary, secondary, and tertiary alcohols [260]. A weak band in this region may be the first overtone of

a $C=O$ absorption or may arise from a small amount of water (unassociated H_2O in CCl_4 absorbs at 3700 cm^{-1} and 3620 cm^{-1}). Organic acids are easily differentiated by their broad irregular absorption, extending to 2000 cm^{-1} or lower.

2. The CH stretching region: 2800–3300 cm^{-1} (3–3.6 μm). The highest of the CH frequencies is exhibited by —C≡C—H at 3300 cm^{-1} (3.03 μm). Aromatic and unsaturated groups show absorptions at 3100 cm^{-1} (3.22–3.34 μm). Highly strained rings also raise the CH frequencies to this range. Most aliphatic compounds absorb at 2800–3000 cm^{-1} (3.34–3.6 μm), and the molecular absorption coefficient is proportional to chain length for straight-chain alkanes [92, 276]. Methoxy or CH_3N groups absorb at 2780–2832 cm^{-1} (3.53–3.60 μm), and —CH_2O— groups often show a weak band at frequencies lower than the ordinary CH stretch. Both aromatic and aliphatic aldehydes have a distinctive absorption at 2700–2775 cm^{-1} (3.6–3.7 μm). A more detailed discussion of the CH stretching region is given elsewhere [56, 268].

3. The "window" region: 1850–2700 cm^{-1} (3.7–5.4 μm). Absorptions lying in this ordinarily clear area are immediately apparent, and their origin is usually easy to deduce. Such bands include amine hydrochloride salts, which show a complex pattern at 2000–2800 cm^{-1} (3.6–5 μm), SH at 2540–2590 cm^{-1} (3.8–3.9 μm), PH at 2275–2440 cm^{-1} (4.1–4.4 μm), —C≡N at 2220–2260 cm^{-1} (4.4–4.5 μm), SiH at 2090–2260 cm^{-1} (4.4–4.8 μm), —C≡C— at 2100–2260 cm^{-1} (4.4–4.8 μm), —N≡C at 2110–2150 cm^{-1} (4.6–4.7 μm), and $C=C=C$ near 1950 cm^{-1} (5.1 μm). Occasionally, Fermi resonance may cause doubling of a band and complicate an otherwise straightforward assignment.

4. The double-bond region: 1430–1950 cm^{-1} (5.1–7 μm). Among the most common and characteristic of the double-bonded groups absorbing in this region are the carbonyls. They are also perhaps the most-studied class of IR-absorbing groups. Whereas certain structures may be differentiated simply from the position of the $C=O$ stretching band, the overlap is such that an unambiguous assignment usually cannot be made without recourse to other regions of the spectrum. As we have already noted, organic acids, and usually aldehydes, are readily identified from their carbonyl band and absorptions in the OH or CH region. Esters, in addition to their $C=O$ stretch, also show a strong C—O—R stretch absorption around 1200 cm^{-1}. Ketones also have bands of medium intensity at 1000–1370 cm^{-1}. Strong absorptions at 1540–1650 cm^{-1} (6.1–6.5 μm) may indicate ionized carbonyl (e.g., metal salts of organic acids), NH in-plane bending in amines, $N=O$ stretch in nitrates, or the $C=O$ stretch of amides. Here again, consideration of other spectral regions is necessary to determine the origin of the absorption. The aliphatic $C=C$ stretch absorptions are found in the range 1630–1690 cm^{-1} (5.9–6.1 μm) unless fluorine is present on one or both carbons. In this case the absorption moves to higher

frequencies, and the number of fluorines may be correlated with the position of the band. Heavier halogens move the band to lower frequencies since, as we saw earlier, the C=C "stretch" involves some CH deformation as well. Valuable structural information can be gained from the position of this band and the associated out-of-plane bending modes around 800–1000 cm⁻¹ (10–12.5 μm) [217]. Aromatic compounds with a low degree of substitution show three (four with good resolution) sharp bands in the 1450–1650-cm⁻¹ range (6–7-μm). Accompanying these are weaker absorptions around 1000–1200 cm⁻¹ (8.3–10 μm) and the characteristic out-of-plane hydrogen deformations at 670–900 cm⁻¹ (11–15 μm). Highly substituted aromatic materials have absorptions of variable intensity near 1400 cm⁻¹ (7.1 μm). The number and position of the substituents can usually be derived from the 670–900 cm⁻¹-pattern or from the overtone and combination band absorptions at 1660–2000 cm⁻¹ (5–6 μm) [274]. Concentrations 10 times normal are required to observe these higher-frequency bands (cf. Fig. A.2.1 in Appendix 2).

5. The fingerprint region: below 1500 cm⁻¹ (6.7 μm). In addition to the characteristic pattern found for individual molecular species in this region, a number of useful group frequencies occur. Among the C—H motions, the methylene scissors vibration is found near 1467 cm⁻¹ (6.82 μm) for alkanes. The methyl asymmetrical CH_3 bending mode falls nearby at 1460 cm⁻¹ (6.85 μm), but the symmetrical methyl bend absorbs near 1380 cm⁻¹ (7.25 μm) and is useful for characterization. It is well known, for example, that *gem*-dimethyl groups give a doublet at this position. Other characteristic CH motions include the out-of-plane hydrogen motions in unsaturated and aromatic compounds. Additional well-known correlations in this region include the CF stretch near 1200 cm⁻¹ (8.3 μm), the C—O—C stretch in ethers and esters around 1200 cm⁻¹, the C—O stretch and OH deformation bands in alcohols at 1000–1260 cm⁻¹ (7.9–10 μm), the P=O stretch near 1200 cm⁻¹ (8.5 μm), the SiO stretch at 1000–1100 cm⁻¹ (9–10 μm), and the CCl stretch vibrations in the vicinity of 700–800 cm⁻¹ (12.5–14 μm).

Every worker in spectroscopy soon learns these and other more specialized correlations relating to his own field of interest, and thus in a few minutes' time is able to form some general opinions about the nature of the unknown material. (A more complete set of correlations is given in Appendixes 1 and 2.)

The next step is to check the reference file for compounds suggested by examination of the spectrum and other evidence. If the unknown is not found immediately, the IR library may be searched using (preferably) a computerized retrieval system (see pp. 59–63) that prints out hits in order of goodness of match. If this approach is not fruitful, critical inspection of the spectra of related structures will often provide clues leading to identification of the unknown, either by direct matching of the spectrum with an authentic sample, or by a reasonable postulate of the structure based on spectral patterns of related compounds [63]. In the latter case, confirming evidence by chemical analysis and/or other methods is desirable and NMR or mass spectroscopy should be used freely or even routinely. A molecular weight determination may give valuable information. In some cases, chemical treatment, such as saponification of an ester, may be necessary. The process described in this paragraph is the critical stage in the identification process and is the point at which the spectroscopist utilizes his background of experience and chemical intuition to deduce a reasonable structure that fits the evidence at hand. This is also the point at which the largest percentage of successful characterizations will be achieved. The process is shown diagrammatically in Fig. 5.20.

An example that occurred as an actual problem in the author's laboratory will illustrate this approach. The spectrum shown in Fig. 5.21 was obtained from a white crystalline solid (melting point ca. 65–70°C), a product of the reaction of styrene, lithium metal, and dichlorodiphenylsilane. It is obviously aromatic; no aliphatic CH, NH, or OH is present. The aromatic is monosubstituted (1600–2000 cm^{-1}) and is connected to an electronegative element in the first row of the periodic chart (688-cm^{-1} and 755-cm^{-1} bands). Biphenyl comes to mind, but comparison with a standard spectrum eliminates this possibility. The C—O—C band is absent (1250 cm^{-1}), so it cannot be phenyl ether. The spectrum of *trans*-stilbene is very similar, but not identical. One thinks next of *cis*-stilbene, but this is a liquid, as is 1,1-diphenylethylene. Tetraphenylethylene melts at 227°C (which is too high). The melting point of triphenylethylene is 72°C, but its spectrum does not match. The spectrum of diphenylacetylene (melting point 62.5°C), however, provides a perfect fit except for the small absorption at 960 cm^{-1}, which is attributed to a *trans*-stilbene impurity. In this case the problem was solved, without recourse to a computerized spectrum retrieval system, simply by checking melting points and a few spectra from the standard file.

Analysis of Mixtures

Spectra of mixtures can sometimes be decoded by using computerized or punched-card file searching techniques as outlined earlier (pp. 59–63). If no more than two or three components are present and their spectra are in the reference file, identification should be possible through the proper use of band and no-band information. Users of interferometer

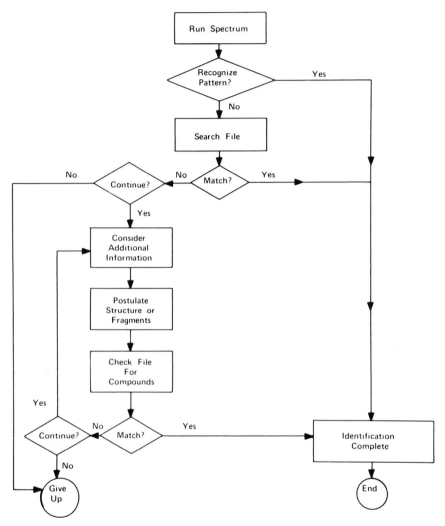

Fig. 5.20 Flow diagram of procedure used by spectroscopist to identify an unknown. Reproduced, with permission, from *American Laboratory* [241]. Copyright International Scientific Communications, Inc.

spectrometers and dispersive spectrometers having computer memory storage of spectra can use a technique known as *absorbance subtraction* or *spectral stripping*, in which reference spectra of the individual components, multiplied by an appropriate scaling factor, are subtracted successively from the spectrum of the mixture, leaving behind the spectra

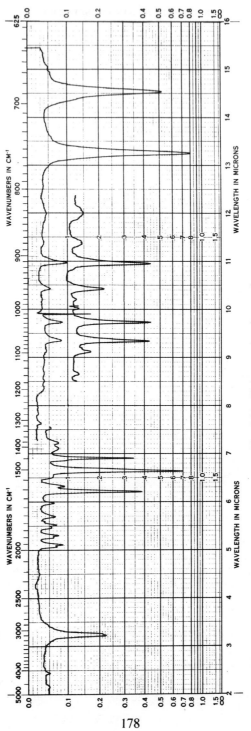

Fig. 5.21 Unknown reaction product.

178

of the remaining components. Double-beam spectrometers can sometimes be used to cancel out bands of a major component optically. The impure sample is contained in the sample beam cell, and a variable-thickness compensating cell containing one known component is placed in the reference beam and the thickness carefully adjusted to cancel the absorptions of the component in the sample [263]. The spectrum is then scanned and theoretically shows only the remaining components of the sample. In practice, molecular interactions, refractive index effects, and Beer's law failure are likely to produce a spectrum with anomalous blips and peculiarly shaped peaks [126]. Also, if the component being canceled has any bands of less than about 25% transmittance, the pen is likely to show little or no response in that region, even though other absorption bands may underlie that absorption (see pp. 40–44). Compensation methods can be quite successful over limited spectral regions, however, especially where the component being canceled has only minor absorptions.

Physical separation of the components, using gas or liquid chromatography, distillation, or solvent extraction, may give the most satisfactory results. Separation theory and techniques have been discussed in detail by Karger, Snyder, et al. [150].

For mixtures of a few components, a partial separation combined with spectral stripping can be used to obtain spectra of the pure components without prior knowledge of their number or identity [125]. The method utilizes the ratio of the spectra from the partially fractionated components. The fractionation procedure can be as simple as partially evaporating the mixture, passing the mixture through a solid absorbant, or partially extracting or precipitating one component. The extent of the separation need not be large, and the effect on the individual materials need not be known. The procedure used in the example is to first measure the spectrum of the mixture (Fig. 5.22a) allow the mixture to evaporate for a while, rescan the spectrum, array ratio the two spectra using a computer, determine the number of flat regions of different heights (numerically equal to the number of components; see Fig. 5.22b); and finally, ascertain mathematically the spectrum of each component. Results with a toluene–cyclohexane–hexane mixture are shown in Fig. 5.23. This procedure is most conveniently done using an interferometer spectrometer with dedicated computer. A general procedure for determining spectra of pure components from spectra of mixtures has been described and applied to polymer analysis by Koenig, D'Esposito, et al. [158].

Use of Correlation Charts

On unidentified components, or in cases in which no headway has been made toward identification, a step-by-step dissection of the spectrum

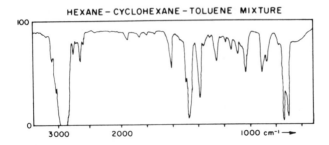

HEXANE – CYCLOHEXANE – TOLUENE MIXTURE

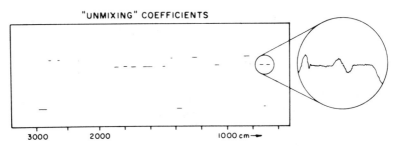

"UNMIXING" COEFFICIENTS

Fig. 5.22 (*a*) Spectrum of toluene–cyclohexane–hexane mixture; (*b*) ratio of absorbances of two successive mixtures, taken after partial evaporation. Reproduced, with permission, from *Analytical Chemistry* [125].

should be made using a correlation chart. The proper function of correlation charts is to suggest compounds and structures for further investigation. *Interpretations based solely on correlations of band positions as obtained from a chart will almost certainly be wrong.*

Correlation charts are a natural outgrowth of the existence of characteristic group absorptions in the IR region. They show the most probable range of absorption frequencies found (determined empirically from studies of a large number of known structures) when a certain group is known to be present. Since the exact position of a group frequency depends on many factors, absorption ranges may be broad and overlap each other considerably.

Properly used, correlation charts can be a valuable aid in characterizing structures. They also can give erroneous and misleading answers if their limitations are not recognized. Thus it is important to keep in mind the following restrictions:

1. Unambiguous structural determinations are usually not possible from correlation charts. Because of overlapping absorption ranges, pertur-

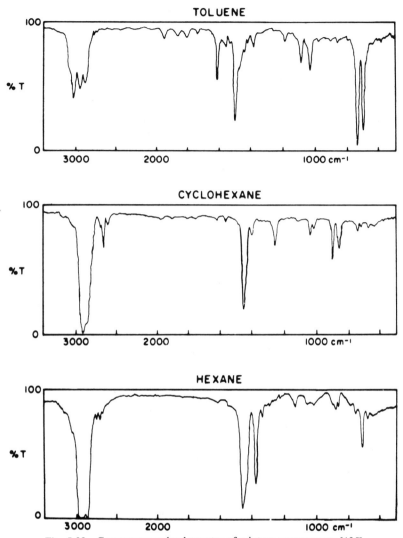

Fig. 5.23 Computer-resolved spectra of mixture components [125].

bations on group frequencies, and ability of charts to show only the strongest and most characteristic bands, the worker who attempts a mechanical interpretation of a spectrum using the chart alone is almost certainly doomed to failure.

2. The spectrum of the material being investigated should be run under approximately the same conditions as were used during compilation

of the chart. To cite an obvious example, most charts are not usable with vapor-state spectra, since many frequencies may fall outside their normal condensed state ranges.

3. The spectrum as a whole, rather than isolated absorptions, should be considered in making an interpretation. Carbonyl groups, for example, show a weak overtone that is easily mistaken for a trace of OH. Strong absorptions in the 690–870-cm^{-1} regions may indicate aromatic groups, but unless accompanied by the sharp absorptions near 1450–1600 cm^{-1} and 1100 cm^{-1}, they are probably due to some other group such as CCl.

4. Band shapes and relative intensities may be of importance equal to or greater than band positions. Aromatic bands are usually sharp, for example, and the relative depths of the 1450–1600-cm^{-1} absorptions may give a clue as to the type of substituent on a phenyl group. The C=C absorption near 1650 cm^{-1} is sharp and not likely to be confused with the amide I band (C=O absorption), which falls at the same place. The practicing spectroscopist will easily recall numerous similar examples.

The proper use of correlation charts, then, is as a guide to suggest possibilities for further investigation. A number of excellent treatments of group frequencies, classified by groups (rather than by wavelength regions), are available [23, 24, 56, 199]. Further study of these monographs plus reference to pertinent spectra in the files and in the literature will soon narrow the possibilities to the point at which the unknown band can be identified with a particular structure, or at least with the presence of a specific group in the molecule.

A correlation chart and group frequency table for some of the more common organic groups are given in Appendixes 1 and 2. More detailed correlation charts have been published [56, 142, 242]. An extensive tabulation of absorption frequencies for many structures has been prepared; effects of solvent–solute interaction and hydrogen bonding on group frequencies are also discussed [75].

Once the major absorptions have been indentified, it is usually not profitable to attempt to interpret minor peaks. No useful purpose is served in analytical work by lengthy conjecture about the origin of weak absorptions. If characterization of minor constituents is deemed necessary, the preferred approach is to separate and identify them individually.

Analytical methods for identifying narrow classes of compounds, based on the use of flow diagrams and the presence or absence of certain bands, are useful for nonspectroscopists. Positive identification *must* be based on direct comparison of the unknown spectrum with an authentic reference spectrum, not on the chart alone. Only those materials considered in the construction of the chart can be identified. Samples must be reasonably pure and sample preparation should be done in the same way as for the reference spectra. Within these limitations, such schemes can be useful and may save considerable searching time.

For the worker who wishes to gain some proficiency in the interpretation of spectra, practice problems have been published [199, 239]. Programmed learning systems are also available [13, 58, 123]. Considerable insight into interpretive infrared spectroscopy may be gained by carefully working through such exercises.

Use of Chemical Derivatives

Whereas the IR spectrum of a compound is a unique fingerprint, it happens not infrequently that certain ambiguities in interpretation arise. The spectra of two compounds, such as adjacent members of a homologous series, may be very similar; the spectral range examined may be too limited to discriminate between similar structures; the spectra may be ill-defined, as for some highly polar molecules; the constituent of interest may not be susceptible to direct determination; or the component sought may be in low concentration and its absorptions virtually obscured by those of a well-banded matrix. In such cases it is often helpful to form a chemical derivative that has a more characteristic spectrum than the original material. A few examples will illustrate.

Highly specific spectra are obtained for barbiturates upon reaction with p-nitrobenzyl chloride [50] or with aqueous copper sulfate–pyridine solutions [172]. Anionic surface-active agents can be identified by precipitating them as barium salts and obtaining their IR patterns [136]. Amino acids can be characterized by the spectra of their 3-phenyl-2-thiohydantoin derivatives [80]. The 3,5-dinitrobenzoates of homologous alcohols show unique spectra [76] and usually can be distinguished even when the melting points of the derivatives are inconclusive. Aliphatic alcohols in fruit flavors have been isolated by paper chromatography and identified by the IR spectra of their p-nitrophenylazobenzoic acid esters [37]. Spectra of 2,4-dinitrophenylhydrazone derivatives of ketones and aldehydes have been reported [140]. Often the X-ray patterns of the compounds are also distinctive and may be used to confirm or resolve the IR result. Even metal ions may be identified from the spectra of their 8-hydroxyquinolates [209] or of other chelating agents [122]. The chemical-derivative approach to both qualitative and quantitative IR analysis would seem to merit further attention from analytical chemists.

Spurious Bands

In identifying unknowns, one frequently finds that a minor, or often a major, component is an unsuspected material that has no relation to the problem at hand. It is not at all unusual, for example, to find spectra, even in the literature, confused by silicone grease from joints or stop-cocks, phthalate ester plasticizers from plastic tubing, or polymeric bottle-cap liners from sample vials. Every laboratory doing IR work can (and should) compile a list of spurious absorptions arising from these and other sources common to that laboratory. A tabulation of commonly occurring spurious peaks has been given by Launer [168], and a portion of his list is given in Table 5.4. Spectra of compounds used to polish alkali halide windows have been published by McCarthy [188].

TABLE 5.4 Common Spurious Absorption Bands

Approximate Frequency (in cm^{-1})	Wave-length (in μm)	Compound or Group	Origin
3700	2.70	H_2O	Water in solvent (thick layers)
3650	2.74	H_2O	Water in some quartz windows
3450	2.9	H_2O	Hydrogen-bonded water; usual in KBr disks
2350	4.26	CO_2	Atmospheric absorption
2330	4.3	CO_2	Dissolved gas from dry ice
2300 and 2150	4.35 and 4.65	CS_2	Leaky cells
1996	5.01	BO_2	Metaborate in the halide window
1400–2000	5–7	H_2O	Atmospheric absorption
1820	5.52	$COCl_2$	Decomposition product in purified $HCCl_3$
1755	5.7	Phthalic anhydride	Decomposition product of phthalate esters or resins
1700–1760	5.7–5.9	C=O	Bottle-cap liners leached by sample
1720	5.8	Phthalates	From plastic tubing
1640	6.1	H_2O	Entrained in sample, or water of crystallization
1520–1620	6.2–6.6	$\left(-C\begin{smallmatrix}O\\O\end{smallmatrix}\right)^{-}$	Reaction product of alkali halide windows or KBr pellet with organic acid

TABLE 5.4 (Continued)

Approximate Frequency (in cm⁻¹)	Wave-length (in μm)	Compound or Group	Origin
1520	6.6	CS_2	Leaky cells
1430	7.0	CO_3^{-2}	Contaminant in halide window
1360	7.38	NO_3^-	Contaminant in halide window
1270	7.9	$SiCH_3$	Silicone oil or grease
1110	9.0	?	Impurity in KBr for disks
1000–1110	9–10	SiOSi	Glass; silicones
980	10.2	K_2SO_4	From double decomp. of sulfates in KBr pellets
935	10.7	$(CH_2O)_x$	Deposit from gaseous formaldehyde
907	11.02	CCl_2F_2	Dissolved Freon-12
837	11.95	$NaNO_3$	(See 1360 cm⁻¹)
823	12.15	KNO_3	From double decomp. of nitrates in KBr pellets
794	12.6	CCl_4 vapor	Leaky cells
788	12.7	CCl_4 liquid	Incomplete drying of cell or contamination
720 and 730	13.7 and 13.9	Polyethylene	
728	13.75	Na_2SiF_6	SiF_4 + NaCl windows
667	14.98	CO_2	Atmosphere
	Any	Fringes	If refractive index of windows is too high, or if a cell is partially empty, interference fringes may appear

Traces of solvent or other contaminants in the *reference beam* will give an inverted spectrum superimposed on the normal sample spectrum. Other spectral artifacts include the Christiansen effect (see pp. 78; 82), energy-level rearrangement due to vibrational perturbations [81], and the "dead pen" found in regions of strong solvent absorption (see pp. 40–44 and 74).

SUMMARY—CHECK LIST FOR UNKNOWN IDENTIFICATION

To optimize acquisition of information from the spectrum, one must minimize extraneous effects from solvent and other interactions. Samples should therefore be run according to well-established and reproducible procedures, preferably as solutions in CCl_4 and CS_2. Insoluble materials may be sampled using films, the mineral oil mull technique, attenuated total reflectance, or the KBr pellet method, as appropriate. It is important to avoid capricious choice of technique for sample preparation.

Identification of unknown materials may be approached as follows.

1. Determine what information is required about the sample and what use will be made of this information.
2. Note obvious characteristics of the material such as form, color, odor, and solubility.
3. Obtain a sample history that is as complete as possible.
4. Run the IR spectrum.
5. Examine the spectrum critically, using knowledge of group frequencies and their variations with molecular structure changes. Use NMR, UV, or mass-spectroscopic data if available.
6. Compare spectra of similar structures suggested by the previous step.
7. If a match is not found, use a spectrum-retrieval system to try to match the unknown, using combined negative and positive band information, as discussed earlier (p. 59).
8. Apply separation procedures if indicated and start again with step (4) on each component.
9. Use correlation charts to suggest additional structures for comparison.
10. Obtain additional data such as elemental analysis; mass, Raman, or NMR spectra; or molecular weights.
11. Postulate a reasonable model for the compound consistent with the spectral data and with its chemical and physical properties.

APPLICATIONS

Polymers

Infrared spectroscopy has been an indispensable aid in the development of the polymer industry. Its use in plastics identification is widely known and practiced. It is of great value in the paint industry for qualitative identification of vehicles, pigments, and solvents; raw materials qualification; quantitation of resin components; curing and oxidation studies; and the diagnosis of product problems. Elastomers, after some physical separation of the components, can usually be identified as to both major and minor components. Fibers and textiles, as well as fiber surface treatments for soil release, fire retardancy, and the like, are often characterized by IR spectroscopy. Beyond such practical uses, it can give valuable information about the tacticity, crystallinity, spacial arrangement, and branching of polymer chains. Indeed, the field of application is so broad that we can do little more than indicate useful areas of application and refer to some of the pertinent literature.

Often, preparation of the sample in a form suitable for analysis is more of a problem than characterizing its spectrum and indeed may be the key to solving the problem being addressed. Some sampling methods that have been employed successfully are detailed in Chapter 4 and in the book by Henniker [118].

Identification

Access to a good collection of reference spectra is essential, and several such compilations are listed in Table 3.3. Identification and analysis of plastics is the subject of an excellent book by Haslam and Willis [115]. Hummel's volumes [131] on IR analysis of polymers, resins, and additives are very useful, as are the text and reference spectra contained in a book authored by members of the Chicago Society for Paint Technology [51].

In many cases, the polymeric material will contain two or more components, in which case the resulting spectrum is approximately the sum of the component homopolymers. With a few exceptions, it is usually impossible to distinguish between copolymers and mixed homopolymers by IR spectroscopy alone. Laminated polymer films give the appearance of being a homogeneous phase in transmission, but if the two sides of the films are examined separately using ATR, spectra of the individual components can often be obtained. In the case of mixtures of copolymers showing mutual interference, the polymer spectra may be simplified by chemical or physical separation of the components [52, 114]. Polyurethanes, for example, can be hydrolyzed with alkali and extracted to fractionate the components for subsequent identification [60]. Uncured paints can be separated by solvent extraction and centrifuging, and the components can be examined using standard sampling techniques.

Cured paints and resins present a more difficult problem, and ATR or pyrolysis techniques can often give a rough idea of the composition of the material. Weathered paints have been identified by grinding the sample and using the KBr disk technique [77, 155]. The diamond window high-pressure cell has been used for forensic microsampling of paints and plastics [259].

Such gross characterizations *can* be misleading, however, since coatings materials are often complex blends of many components, some of which may absorb strongly and others weakly. Craver [64] shows an example of two different phthalic alkyd resins in which the spectra are almost completely dominated by only 26% of the phthalate moiety. A more complete analysis of such compositions requires careful separation of the components, with characterization of each of the fractions. Systematic procedures for analyses of resin-based coatings are given by Kappelmeier [149], who includes methods for spot tests, saponification, and other chemical procedures as well as solvent extraction, such as are illustrated by the flow diagram for analysis of polyesters in Fig. 5.24.

Pigments can often be identified from their IR spectra [2, 131] or by X-ray diffraction, perhaps combined with metals analysis.

Coatings analysis is often complicated by the fact that some ingredients

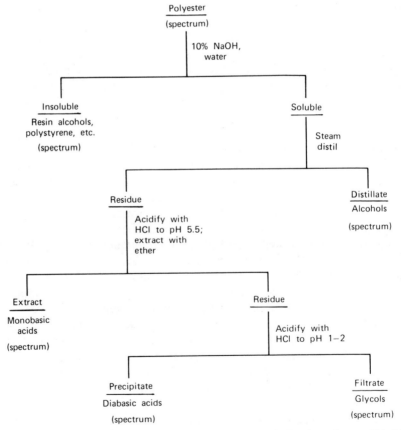

Fig. 5.24 Analytical scheme for chemical and physical analysis of a polyester [64, 149]. Reproduced with permission of Wiley-Interscience.

are themselves complex mixtures, such as shellac, rosin, wax, and natural resins. Any separation procedure ideally should isolate such mixtures as a whole from other similarly complex components. In characterizing such materials, reference spectra are useful but are often ambiguous. A knowledge of compatible formulations and their expected components can be extremely useful. Such formulations can be found in several works [189, 198, 249].

Many pitfalls in both sampling and interpretation await the naive, hasty, or careless analyst. Some of these are identified by Craver [64], whose chapter on analysis of surface coatings should be read by anyone involved with this type of analysis. Practical applications of IR and GC in the paint industry are discussed by Helmen [117], who also gives illustrative examples.

Elastomers are usually complex mixtures containing one or more primary polymers, pigments and fillers, plasticizers, accelerators, antioxidants, stabilizers, lubricants, antistatic agents, and so on. Identification of all these components by IR examination of the gross elastomer is highly unlikely. In fact, an elastomer loaded with carbon black may be so opaque that it will give no spectrum at all; thus separation is necessary. Analyses of specific polymer systems are dealt with in various publications; some of these are reviewed and referenced by Hampton in his comprehensive review of applied IR spectroscopy in the rubber industry [114]. Solvent extraction and chromatographic methods, including thin-layer, gel-permeation, column, and gas chromatography are often used to separate components for IR analysis.

Fibers and textiles are conveniently sampled using ATR [132], although they can also be ground and pressed into KBr disks [187], pyrolyzed [49], or fragmented and pressed into thin, unsupported sheets [214]. Fibers that cannot be ground or formed into films can be held in a parallel configuration in a microholder that permits recording of useable spectra on only four to six fibers [48]. Chemically modified cottons have been studied using difference spectroscopy [187, 203]. If the fabric is treated with hydrophobic or olephobic finishes or a flame-retardant finish, the ATR or transmission spectrum may show both the treatment and the fiber. Often the treatment can be separated and identified by a simple solvent extraction using chloroform, tetrachlorethylene, or methanol [170]. The solvent may be evaporated on an ATR plate, and only miniscule quantities of extract are needed to give a recognizable spectrum (see pp. 86–95). More complex extraction procedures are used when flame retardants, lubricants, and softeners are all present on the fiber [195]. A catalog of spectra for the identification of chemical finishing agents has been compiled by O'Connor, McCall, et al. [204].

Structure

Considerable effort has been expended in the study of polymer structures by IR, in many cases with considerable success. Crystalline polymers usually give different IR spectra than amorphous structures. The arrangement of monomer units in the polymer and their spacial configuration and the arrangement, packing, and branching of the polymer chains can all be studied by IR spectroscopy. It is sometimes possible to distinguish between random and block copolymers, for instance, if one of the monomer units contains associative groups and the other does not. In this case the amount of association gives a measure of polymer randomness. In some cases one of the monomer units is sensitive to its environment and shows a frequency change in a copolymer as opposed to the homopolymers. An example has been cited [216] of the vinyl chloride–vinylidene chloride system, in which the 1250-cm^{-1} (8-μm) band of pure poly(vinyl chloride) shifts to 1203 cm^{-1} (8.3 μm) in the copolymer spectrum. The latter absorption is presumably due to isolated (—CH$_2$CHCl—) units in the chains of poly(vinylidene chloride).

Crystallinity is usually studied with the aid of polarized IR radiation [35]. Some polymers exhibit "crystalline" and "amorphous" bands at different frequencies. Polyethylene terephthalate, for example, has a band at 1343 cm^{-1} attributed to crystalline structure and one at 1370 cm^{-1} arising from amorphous regions. The ratio of the two forms can be calculated from these absorptions [79]. Krimm [163] has studied the polymer-chain organization in crystalline polyethylene by IR.

Spectra of isotactic and atactic polystyrene are shown in Fig. 5.25. Differences are not pronounced but can be seen in the 1060-cm^{-1} region. Polymorphism in polybutene-1 is illustrated in Fig. 5.26, where modification I has a threefold helical conformation, modification II has a fourfold helical conformation, and modification III has an orthorhombic structure [182]. Sequence distribution in ethylene–propylene copolymers has been determined using IR [258].

Absorbance subtraction (see pp. 171–181) with an interferometer spectrometer has been used [157] to obtain the spectrum of crystalline 1,4-transpolychloroprene from spectra of the partially crystalline material and the amorphous polymer. Identical results are obtained with a computerized dispersive spectrometer (Fig. 5.27).

Reactions

Absorbance subtraction was used by Koenig [157] to study surface oxidation of polybutadiene at 30°C, and results are shown in Fig. 5.28. A change in the ratio of *cis* and *trans* unsaturation can be detected after

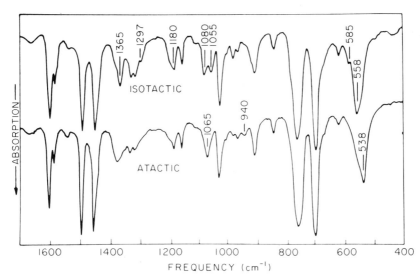

Fig. 5.25 Infrared spectra of isotactic and atactic polystyrene. Reproduced, with permission, from *Applied Spectroscopy* [182].

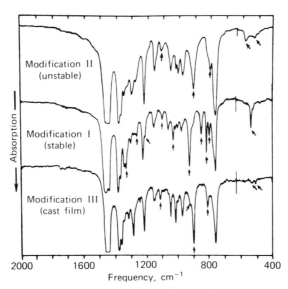

Fig. 5.26 Infrared spectra of three polymorphic forms of isotactic polybutene-1. Reproduced, with permission, from *Applied Spectroscopy* [182].

191

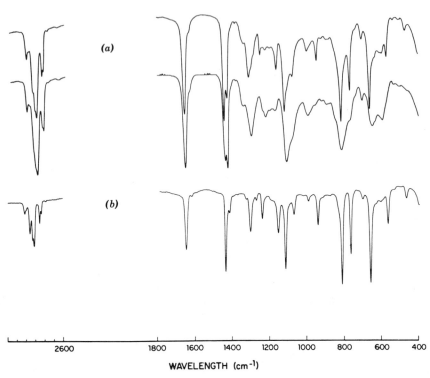

Fig. 5.27 (*a*) Upper spectrum: room-temperature spectrum of naturally occurring mixture of crystalline and amorphous 1,4-*trans*polychloroprene; lower spectrum: −80°C spectrum of 1,4-*trans*polychloropropene, amorphous component only. (*b*) Computer-calculated difference spectrum for crystalline component of 1,4-*trans*polychloroprene. Courtesy Perkin-Elmer Corporation.

only 10 h (3000 cm^{-1} and 975 cm^{-1}). Some oxidation (formation of C—O) is suggested by the 1065-cm^{-1} band. After longer exposures, oxidation gives rise to absorptions from OH (3300 cm^{-1}) and C=O (1700 cm^{-1}, 1720 cm^{-1}, and 1770 cm^{-1}). Irradiation damage to polyethylene has been studied in a similar manner [250]. Weathering damage to acrylonitrile–butadiene–styrene terpolymer films was followed using IR [66]. Ultraviolet-induced degradation of polycarbonate plastic was evaluated using ATR; depth profiling was done by physically removing layers of surface [99]. The ATR technique was used to study ozone degradation of elastomers [7].

The methods used in deducing structural information from vibrational spectra are beyond the scope of this chapter; for more detailed information the reader is referred to specialized monographs [135, 275] and

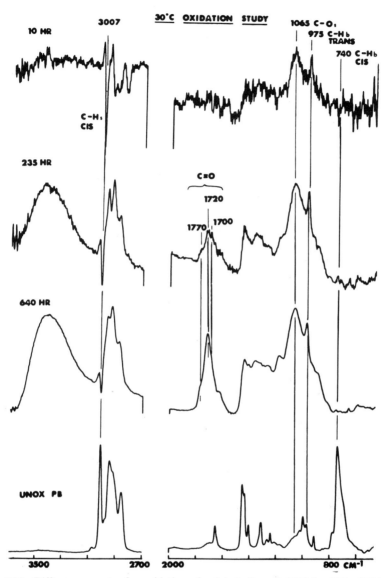

Fig. 5.28 Difference spectra for oxidation of polybutadiene at 30°C for 10 h, 235 h, and 640 hr. Reproduced, with permission, from *Applied Spectroscopy* [157].

other literature [36, 127] and also to the Applications Reviews on polymers that appear in odd-numbered years in the April review issue of *Analytical Chemistry*.

Surfactants

The analytical problems involved in isolating and identifying surfactants can be very challenging indeed. These materials are often minor components in formulations; they are usually used in combination; and even in their "pure" state they are generally mixtures of related structures, often contaminated with starting materials, reaction by-products, and nonsurfactant additives. The first step, then, will probably involve isolating the surfactants from the other materials. After separating the mixed surfactants, one can begin attempts at specific identification.

Systematic separation and analysis of surfactants has been discussed in detail by Rosen and Goldsmith [227], who recommend suitable chemical and physical analyses, including IR, for characterizing the products of the separation. A monograph by Hummel [130] includes a number of IR spectra.

Biological Systems

Because biological materials often do not give well-defined IR spectra (at least at room temperature), the use of IR has not become as popular among biologists as among chemists. Nevertheless, some interesting applications have been reported. One of these involves the use of ATR to sample the surface of living skin. The uptake of both inert and bioactive materials can be traced as a function of time [90, 222]. Infrared spectroscopy has also been put to more conventional uses, such as analysis of blood for lipids [94, 231] and differentiation of tissue types [34, 70]. Applications of IR to biological systems have been reviewed [139, 206, 207, 252].

Inorganics, Metalorganics, and Coordination Compounds

Although Coblentz examined the IR-transmission spectra of a great many inorganic substances in 1906–1908, the potential of the IR technique for qualitative, quantitative, and structural analysis of inorganic materials was largely overlooked until the 1950s. During this decade several compilations of spectra appeared, which pointed out the usefulness of IR for identification, particularly in conjunction with X-ray and emission analysis. In addition to the more traditional nonorganics such as UF_6, $SiCl_4$,

BF$_3$, and NH$_3$, metal coordination compounds have been extensively studied in recent years. Spectral data and references pertinent to these two classes of materials are found in several monographs [87, 109, 186, 200]. Other related works include books on inorganic vibrational spectroscopy [141] and IR and Raman spectroscopy of lunar and terrestrial minerals [151]. Low-frequency (far-IR and Raman) vibrations of inorganic and coordination compounds are discussed by Ferraro [87]. Soils and their constituents have been characterized using IR spectroscopy [82, 83, 89, 138, 261], as have cement minerals [101].

Some classes of hybrid molecules to which vibrational spectroscopy has been successfully applied include organosilicon [6] and organophosphorus [202, 253] compounds. Vibrational spectra of transition-element compounds have been reviewed [104]; studies on organometallics are summarized by Adams [1] and Goldstein [104].

Environmental Problems

It is natural that a technique that can give both positive identification and a quantitative measure of pollutants would be widely adopted for environmental monitoring. A growing body of literature on the application of IR spectroscopy to environmental problems testifies to this fact.

Air analysis can be done in the field with simple portable spectrometers [103, 269, 270]. Long-path (10–20-m) gas cells are usually employed, and sensitivity for many pollutants is in the low-parts-per-million range. Greater discrimination between interfering absorbers and better sensitivity can be achieved with the use of high-resolution spectrometers. A useful set of 600 spectra of hazardous gases and vapors and their physical properties has been compiled by Thompson [254]. Quantitative analysis of gaseous pollutants is discussed later (pp. 259–260).

Infrared has also been used to study atmospheric reactions (e.g., photolysis and oxidation) using a special long-path cell in which the contents could be irradiated with simulated sunlight and the absorption spectrum of the reactants and products monitored continuously [98, 213]. Typical spectra taken before and after 100 min of radiation are shown in Fig. 5.29. Concentration dependence on irradiation time is given in Fig. 5.30. Detection limits for a number of smog components, studied in a cell of 1-km path length, are given in Table 5.5.

Reactions of the herbicide Linuron in different soil systems have been studied using Fourier transform IR [185]. Much larger amounts of herbicide are required on peat soils to be effective. Absorbance subtraction was used to detect a chemical reaction of Linuron absorbed on peat particles (but not on clay).

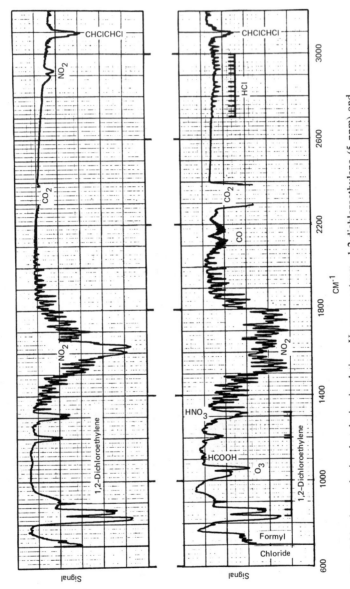

Fig. 5.29 Atmospheric photolysis simulation. Upper spectrum, 1,2-dichloroethylene (5 ppm) and NO_2 (1 ppm) before irradiation; lower spectrum, after irradiation. A 500-m optical path was used. Note that abcissa scale is reversed from customary presentation. Reproduced, with permission, from *Environmental Science and Technology* [98].

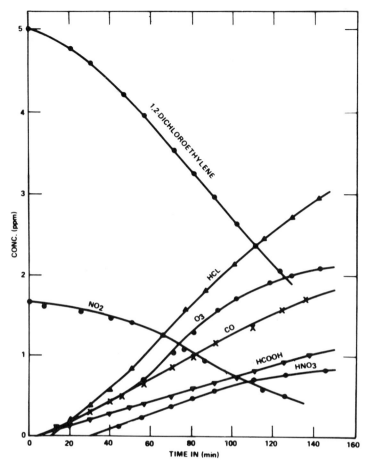

Fig. 5.30 Photooxidation of 1,2-dichloroethylene and NO₂ [98].

There have been many attempts to fingerprint oils from various sources so that the source of oil spills can be identified. The major problem is that the evaporation of lighter components and weathering of the residue changes the absorption pattern of the oil as a function of time (Fig. 5.31). Also, contaminants such as animal or vegetable oils may confuse the spectrum. The most successful approach to infrared characterization of weathered oils has been to measure the absorbance at a number of wavelengths, selected statistically to show maximum sensitivity to the source of the oil and minimum sensitivity to weathering [3, 38, 153, 183]. Fresh oil samples can be artificially weathered to match spill samples;

TABLE 5.5   Detection Limits for Air Pollutants, 1-km Optical Path[a]

Compound	Measurement Frequency (in cm^{-1})	Detection Limit (in ppb)
Ammonia	931.	2
	967.5	2
Formaldehyde	2779.	4
	2781.5	
Formic acid	1105.3	4
Nitric acid	896.	4
Nitrous acid (cis)	853.	4
Peroxynitric acid	803.	4
Nitrogen pentoxide	740.	2
	1248.	2
Ozone	1055.	10
Peroxyacetyl nitrate	793.	4
	1162.	3
Hydrogen peroxide	1250.	8
Methanol	1033.	2
Methyl nitrate	853.	2

[a] Data from Pitts and Finlayson-Pitts [213].

water temperature appears to have the predominant role in affecting the rate, although illumination and agitation are also factors [3, 12].

Analysis of Remote Atmospheres and Objects

Sampling of remote terrestial and celestial atmospheres or objects for IR characterization presents some difficult problems. First, the energy limitations discussed earlier mean that obtaining a usable signal is a major challange. The sun can sometimes be used as a source but has some obvious limitations. Use of controlled remote IR sources or retroreflectors is possible only occasionally. Emission spectra from warm gases, such as stack emissions, can be observed by correlation spectroscopy (see pp. 259–260). Tunable lasers seem to offer the most promise for remote air monitoring [124].

The second problem is that the intense atmospheric H_2O and CO_2 absorptions, especially at long paths (≥ 1 km), require that one work in the clear windows between bands (Fig. 5.32).

Surely something close to the ultimate in remote IR analysis is described by Larson and Fink [166], who combined the techniques of

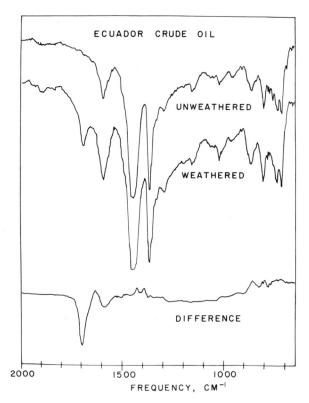

Fig. 5.31 Infrared spectra of unweathered and weathered crude oil and optical difference spectrum (obtained with weathered oil in sample beam and unweathered oil in reference beam). Reproduced, with permission, from *American Laboratory* [38]. Copyright International Scientific Communications.

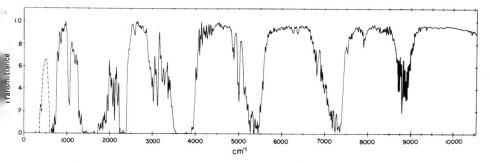

Fig. 5.32 Infrared transmission spectrum of the earth's atmosphere in 1-10 μm region. Major atmospheric windows occur around 1000 cm^{-1}, 2000 cm^{-1}, 2800 cm^{-1}, 4500 cm^{-1}, 6200 cm^{-1}, and 8200 cm^{-1}. Adapted from Ridgway, Larson, et al. [226] and reproduced, with permission, from *Applied Spectroscopy* [166].

Fourier transform spectroscopy with wide-aperture telescopes to characterize the surface composition of small-solar-system objects such as asteroids and satellites. Reflectance spectra were observed, with the sun as the source. An optical arrangement was used in which the signal from a selected sampling area, as viewed through the telescope, was compared interferometrically to a signal received from an adjacent area of sky. Thus the effect of high background radiation levels was canceled so that the weak signal from the target could be detected by signal averaging many thousands of scans.

As might be anticipated, frosts of different composition show quite different IR spectra (Figs. 5.33 and 5.34). In addition, H_2O ice shows a band at 6056 cm^{-1} whose intensity is temperature dependent. Thus the temperature of remote ice objects can be estimated from their reflectance spectra. Many minerals have characteristic reflection spectra that permit at least partial identification of the composition of asteroids and planets (Fig. 5.35).

FREQUENCY ASSIGNMENTS

Whereas a strictly mechanistic correlation of band positions with the presence of certain molecular groups can be of some value, such a dogmatic approach has limited use. Group frequencies can be used more intelligently when the motions of the atoms in the group (bond stretching, bending, and rocking) have been identified, not only to characterize specific groups, but also to predict possible shifts or anomalous effects resulting from adjacent chemical groups. Frequency assignments are essential, of course, for thermodynamic calculations based on spectral data. Although it is not always easy to assign the origin of a group frequency, especially when mixed vibrations are involved, a number of methods have been applied to the problem. They are discussed briefly in the following paragraphs.

Comparison with Related Structures

Undoubtedly most group-frequency correlations are carried out by comparing a large number of related (and preferably simple) structures. A change of one atom or group at a time produces a corresponding change in the spectrum.

Successful application of this method depends on the availability of suitable reference materials. The spectra of polymers often can be interpreted by comparing the spectra of monomer units or of model compounds that contain some of the same isomeric structures as might be

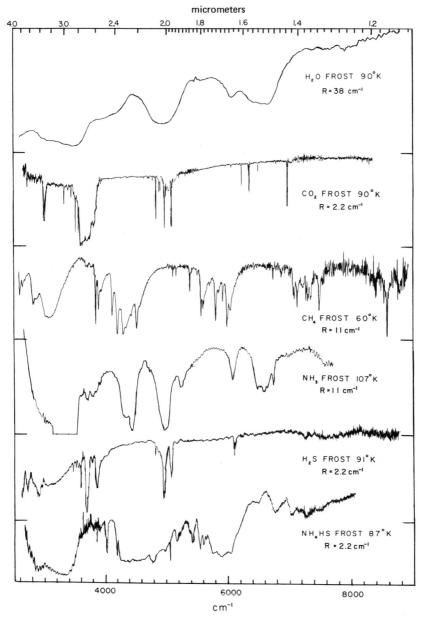

Fig. 5.33 Infrared reflectance spectra of several laboratory frosts. Reproduced, with permission, from *Applied Spectroscopy* [166].

201

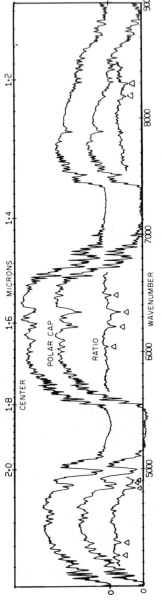

Fig. 5.34 Spectra of center of planet Mars, polar cap, and their ratio. Triangles indicate spectral features unique to polar cap and compatible with presence of CO_2 ice. Reproduced, with permission, from *Astrophys. J. Lett.* [167].

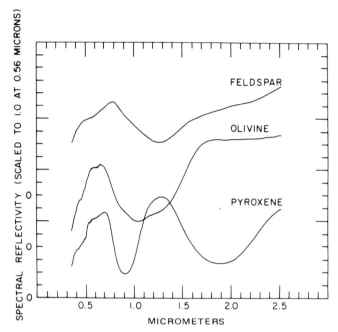

Fig. 5.35 Spectra of the three basic rock-forming silicate mineral groups. Ferrous iron is responsible for all prominent absorptions seen here. Adapted from Gaffey [97] and reproduced, with permission, from *Applied Spectroscopy* [166].

anticipated in the polymers. Thus the structure of polyvinyl chloride [161] has been determined with the help of model compounds such as $CH_3CH_2CH_2Cl$, $CH_3CH_2CHClCH_3$, and $CH_3CH_2C(CH_3)_2Cl$. It was found possible to assign rather restricted frequency ranges to the C—Cl absorptions, depending on whether the Cl is *trans* to a H atom or to a C atom across the common C—C bond [237]. These data together with results of isotopic substitution and a symmetry analysis on the unit cell provided evidence to confirm the polymer configuration postulated from X-ray-diffraction studies. Tables of frequencies and assignments for a large number of simple molecules, compiled and evaluated by Shimanouchi [235], often provide useful data on model structures.

The correlation method has been carried a step farther by Colthup, who has calculated electron densities in substituted ethylenes [57] and aromatic compounds [55] using CNDO/2 molecular orbital theory. He finds a number of simple relationships between various frequencies and electron density.

Intensity Measurements

An obvious extension of group frequencies is the concept of group intensities. This parameter has been explored by a number of workers, but, as we have seen, the experimental difficulties are somewhat greater in intensity spectroscopy than in frequency spectroscopy. Nevertheless, a number of interesting and useful relationships have been advanced. Lippert and Mecke [177] have suggested a two-dimensional correlation diagram on which frequency is plotted as the abscissa and intensity as the ordinate. This type of plot has been used [255] to differentiate the various carbonyl absorptions. Band intensities in various —C≡N,

$\diagdown \diagup$
$C{=}C$, —N=O, and —S=O containing compounds have also been
$\diagup \diagdown$

studied in KBr disks [91]. Successful use of the intensity scale obviously demands a high-resolution spectrometer and the accumulation of considerable background information on a large number of compounds.

This method is not applicable to certain groups that show large variations in intensity (e.g., because of mesomeric effects). The —C≡N absorption, for example, varies in intensity by a factor of 10 or more from compound to compound. Fortunately, it falls in a region of the spectrum where it is seldom confused with other absorptions.

Raman and Other Spectroscopic Data

The principles and uses of Raman spectroscopy are discussed elsewhere [56]. It should be pointed out here, however, that for more symmetrical molecules especially, Raman activity, intensities, and depolarization ratios can be of immense value in distinguishing totally symmetric from nonsymmetric vibrations. The uses of Raman spectroscopy as an aid in determination of molecular symmetry are well known, and no serious study of molecular structure can be attempted without Raman data.

Other spectroscopic techniques, such as UV spectroscopy, used principally in the study of aromatic, unsaturated, and other species containing π-electrons or unshared electron pairs, and microwave spectroscopy, which permits attainment of accurate molecular dimensions by the detailed study of rotational spectra, can at times offer valuable supporting data for frequency assignments.

Rotation–Vibration Band Structure

It was shown as early as 1933 that the shape and separation of the rotational wings in vapor spectra are determined by the direction of change of electric moment during a vibration and the moments of inertia of the molecule [100]. This work, which originally applied to symmetrical molecules, was subsequently extended to include unsymmetrical structures [11, 128]. Thus, knowing the molecular configuration, one can often predict the shape of the rotational envelope for a particular vibration and calculate the separation of the P–R maxima. Only infrequently, however, can molecular symmetry be deduced from band contours alone. High-resolution studies of rotational structure give more detailed information [4].

By Calculation of Force Constants

Force constants may be calculated with any degree of sophistication ranging from the assumption of a diatomic-type vibration for the group in question to a complete solution for all force constants in the molecule, including nonbonded atom interactions. The former method, although simple, is too crude to give results of much value, and the latter is too complex for most practical problems. As discussed earlier, several more realistic force fields have been applied to obtaining force constants in complicated molecules, and considerable progress has been made. The point of interest is that force constants, when properly calculated and used, may be transferred from one molecule to another and frequencies actually calculated with considerable accuracy. Another benefit is that the normal coordinates and potential energy distribution give the actual form of the atom displacements and clearly show the presence of mixed vibrations.

For a more detailed discussion of these methods, the reader is referred to the literature [46, 56, 78, 88, 272, 273].

Isotopic Substitution

We have already seen (pp. 142–143) how substitution of deuterium for hydrogen shifts the corresponding absorption bands by a factor of 1.3–1.5 (ca. $\sqrt{2}$) and how the shifts for different vibrational modes differ slightly. Deuterium substitution is often relatively easy to accomplish and has been a popular method for studying vibrations involving hydrogen.

Other isotopic species, however, also provide useful information, and in some cases natural abundance is sufficient to allow definitive studies to be made. Decius, for example, in a series of alkaline earth carbonates [72], found absorptions from $(C^{13}O_3)^{-2}$ at the natural abundance of 1.1% C^{13}. Bands due to $(N^{15}O_3)^-$ were also observed in spectra of KNO_3 [71]. West and Glaze [265], using Li^6CH_3 and Li^7CH_3, showed that the compound is polymeric even in the vapor state and assigned the Li—C vibrations to bands in the 340–570-cm^{-1} region. Metal isotope substitution as a method of aiding frequency assignments is discussed by Hutchinson, Eversdyk, et al. [133].

As a final example of the use of isotopic substitution in frequency assignment, we refer to the work of Linton and Nixon on silyl cyanide [175, 176]. Previous authors, reasoning by analogy and by chemical properties, had assumed the product of the reaction of H_3SiI and AgCN to be silyl isocyanide. Linton and Nixon, however, suspecting that the normal cyanide was formed in the reaction, synthesized materials enriched with C^{13} and N^{15}. The calculated shifts for the normal cyanide stretch frequency were −51 cm^{-1} for $C^{13}N$ and −31 cm^{-1} for CN^{15}. For the isocyanide, the calculated displacements were −43 cm^{-1} for —NC^{13} and −38 cm^{-1} for $N^{15}C$. The observed differences were −51 cm^{-1} for the C^{13}-enriched material and −30 cm^{-1} for the N^{15}-containing sample, providing strong evidence for the normal cyanide rather than the isocyanide structure.

Frequencies in pairs of isotopically related molecules can be related quantitatively through the Teller–Redlich product rule and through various sum rules [272]. Pinchas and Laulicht [212] have compiled much data on the IR spectra of isotopically substituted compounds.

Polarization Data on Oriented Samples

The polarization technique has been particularly valuable in studies of oriented polymers and single crystals. An absorption band in such a specimen shows maximum depth when the transition moment is parallel to the electric vector of the radiation and no absorption when it is perpendicular. One must, however, be careful to distinguish between transition moment and bond direction since they do not necessarily coincide.

Polarization measurements can thus be used in two ways: (1) to relate transition moments to molecular orientation in crystals of known structure and (2) to determine structure of crystals or polymers from measurements of dichroic ratios. In one such study, polarized ATR data on single-crystal stearic acid and stearic acid d_{35} have been used to deduce vibrational assignments [196]. For further discussion of the polarization method, the literature should be consulted [35, 161].

Frequency Shifts in Solvent

We have seen (pp. 164–167) that the various vibrational modes react differently to changes in solvent. This effect is the basis of an ingenious method of group-frequency characterization in which the shift of the unknown band in various solvents is compared with the displacement of an analogous known frequency in a different molecule. Bonds of the same type (e.g., carbonyl groups) show a similar pattern of frequency shifts in a wide variety of solvents, and this effect has been used to distinguish the C=O band in pyridones from the C=C absorption [29]. Figure 5.36 shows the type of plot that resulted from this study. The smallest shifts are found with relatively inert solvents such as heptane and cyclohexane, and the largest displacements with hydrogen-bonding solvents such as bromoform and methylene iodine. Another example of group-frequency characterization is that of the C=S absorption, the displacement of which has been plotted against the shift of C=O in acetophenone [28]. This work clearly resolved the anomalies in the C=S group-frequency assignment.

The method is not applicable to nonpolar bonds. Also, situations will occasionally be found in which the solvent will associate preferentially with another portion of the molecule, so that the expected shift is not observed in the frequency being studied. Despite these limitations, however, the solvent-shift technique has much potential in the identification of group frequencies and elucidation of structure.

In Chapter 1 it was stated that optical isomers cannot be distinguished by IR spectroscopy. This statement is not quite true; it has been found that use of an optically active solvent with an interferometer spectrometer gives small but measurable changes in band positions and intensities,

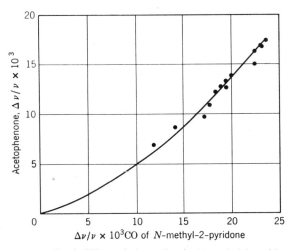

Fig. 5.36 Frequency shift of 1700-cm^{-1} absorption in N-methyl-2-pyridone. Reproduced, with permission, from *Spectrochimica Acta* [29].

which differ for the two isomers and can be detected by absorbance subtraction [110]. The procedure was illustrated by measurements on optically active malic acid (M) in the enantiomorphs of 2-octanol (O). Surprisingly, the spectra of +M in −O was quite different from that of −M in +O; spectra of −M in −O and +M in +O differed from each other and from those of the other two pairs. The explanation for the phenomenon may lie in the fact that the Michaelson interferometer has a weak residual elliptical polarization, which would, of course, remove the spectral degeneracy of the stereoisomers.

References

1. Adams, D. M., *Spectrosc. Prop. Inorg. Organomet. Comp.,* **5,** 239 (1972).

2. Afremow, L. C., and J. T. Vandeberg, *J. Paint Technol.,* **38,** 169 (1966).

3. Ahmadjian, M., C. D. Baer, P. F. Lynch, and C. W. Brown, *Environ. Sci. Technol.,* **10,** 777 (1976).

4. Allen, H. C., Jr., and P. C. Cross, *Molecular Vib-Rotors,* Wiley, New York, 1963.

5. Altshuller, A. P., and A. F. Wartburg, *Appl. Spectrosc.,* **15,** 67 (1961).

6. Anderson, D. R., Infrared, Raman and Ultraviolet Spectroscopy, in A. L. Smith, Ed., *Analysis of Silicones,* Wiley, New York, 1974, Chap. 10, pp. 247–286.

7. Andries, J. C., and H. E. Diem, *J. Polym. Sci., Polym. Lett. Ed.,* **12,** 281 (1974).

8. Arnold, J., J. E. Bertie, and D. J. Millen, *Proc. Chem. Soc. (London),* **1961,** 121 (1961).

9. Badger, R. M., *J. Chem. Phys.,* **2,** 128 (1934); **3,** 710 (1935).

10. Badger, R. M., and S. H. Bauer, *J. Chem. Phys.,* **5,** 839 (1937).

11. Badger, R. M., and L. R. Zumwalt, *J. Chem. Phys.,* **6,** 711 (1938).

12. Baer, C. D., and C. W. Brown, *Appl. Spectrosc.,* **31,** 524 (1977).

13. Baker, A. J., T. Cairns, G. Eglinton, and F. J. Preston, *More Spectroscopic Problems in Organic Chemistry,* 2nd ed., Heyden, London, 1975.

14. Baker, A. W., and A. T. Shulgin, *J. Am. Chem. Soc.,* **80,** 5358 (1958).

15. Barrachough, C. G., D. C. Bradley, J. Lewis, and I. M. Thomas, *J. Chem. Soc.,* **1961,** 2601 (1961).

16. Barrow, G. M., *Spectrochim. Acta,* **16,** 799 (1960).

17. Bauer, E., and M. Magat, *J. Phys. Radium,* **9,** 319 (1938).

18. Becker, E. D., *Spectrochim. Acta,* **17,** 436 (1961).

19. Bell, J. V., J. Heisler, H. Tannenbaum, and J. Goldenson, *J. Am. Chem. Soc.,* **76,** 5185 (1954).

20. Bellamy, L. J., *J. Chem. Soc.,* **1955,** 2818 (1955).

21. Bellamy, L. J., *Appl. Spectrosc.*, **16**, 61 (1962).

22. Bellamy, L. J., The Origins of Group Frequency Shifts, in M. J. Wells, Ed., *Spectroscopy*, The Institute of Petroleum, London, 1962, pp. 205-221.

23. Bellamy, L. J., *Advances in Infrared Group Frequencies*, Barnes and Noble, New York, 1968.

24. Bellamy, L. J., *The Infrared Spectra of Complex Molecules*, Vol. 1, 3rd ed., Wiley, New York, 1975.

25. Bellamy, L. J., and H. E. Hallam, *Transact. Faraday Soc.*, **55**, 220 (1959).

26. Bellamy, L. J., and A. J. Owen, *Spectrochim. Acta*, **25A**, 329 (1969).

27. Bellamy, L. J., and R. J. Pace, *Spectrochim. Acta*, **25A**, 319 (1969).

28. Bellamy, L. J., and P. E. Rogasch, *J. Chem. Soc.*, **1960**, 2218 (1960).

29. Bellamy, L. J., and P. E. Rogasch, *Spectrochim. Acta*, **16**, 30 (1960).

30. Bellamy, L. J., and R. L. Williams, *Transact. Faraday Soc.*, **55**, 14 (1959).

31. Bent, H. A., *Chem. Rev.*, **61**, 275 (1961).

32. Bernstein, H. J., *Spectrochim. Acta*, **18**, 161 (1962).

33. Bertran, J. F., L. Ballester, L. Dobrihalova, N. Sánchez, and R. Arrieta, *Spectrochim. Acta*, **24A**, 1765 (1968).

34. Bessette, F., *Can. J. Spectrosc.*, **20**, 126 (1975).

35. Bluhm, A. L., Polarized Infrared Spectroscopy, in G. L. Clark, Ed., *Encyclopedia of Spectroscopy*, Reinhold, New York, 1960, pp. 522-531.

36. Boerio, F. J., and J. L. Koenig, *J. Macromolec. Sci., Rev. Macromolec. Chem.*, **C7**, 209 (1972).

37. Bosvik, R., K. V. Knutsen, and C. F. E. von Sydow, *Anal. Chem.*, **33**, 1162 (1961).

38. Brown, C. W., P. F. Lynch, M. Ahmadjian, and C. D. Baer, *Am. Lab.*, **7**(12), 59 (1975).

39. Brown, T. L., *Spectrochim. Acta*, **10**, 149 (1957).

40. Brown, T. L., *Chem. Rev.*, **58**, 581 (1958).

41. Brownlee, R. T. C., J. Di Stefano, and R. D. Topsom, *Spectrochim. Acta*, **31A**, 1685 (1975).

42. Brownlee, R. T. C., and R. D. Topsom, *Spectrochim. Acta*, **31A**, 1677 (1975).

43. Bryson, A., *J. Am. Chem. Soc.*, **82**, 4858, 4862 (1960).

44. Buckingham, A. D., *Proc. Roy. Soc. (London)*, **A248**, 169 (1958); **A255**, 32 (1960).

45. Buckingham, A. D., *Transact. Faraday Soc.*, **56**, 753 (1960).

46. Califano, S., *Vibrational States*, Wiley, New York, 1976.

47. Cannon, C. G., *Spectrochim. Acta*, **10**, 341 (1958).

48. Carlsson, D. J., T. Suprunchuk, and D. M. Wiles, *Text. Res. J.*, **47**, 456 (1977).

49. Cassels, J. W., *Appl. Spectrosc.*, **22**, 477 (1968).

50. Chatten, L. G., and L. Levi, *Appl. Spectrosc.*, **11**, 177 (1957).

51. Chicago Society for Paint Technology, *Infrared Spectroscopy: Its Use in the Coatings Industry*, Federat. Soc. Paint Technol., Philadelphia, 1969.

52. Cobler, J. G., Plastics, in F. J. Welcher, Ed., *Standard Methods of Chemical Analysis*, Vol. IIB, 6th ed., Van Nostrand, New York, 1963, pp. 2034–2104.

53. Coggeshall, N. D., and E. L. Saier, *J. Chem. Phys.*, **15**, 65 (1947).

54. Colthup, N. B., *J. Chem. Educ.*, **38**, 394 (1961).

55. Colthup, N. B., *Appl. Spectrosc.*, **30**, 589 (1976).

56. Colthup, N. B., L. H. Daly, and S. E. Wiberley, *Introduction to Infrared and Raman Spectroscopy*, 2nd ed., Academic Press, New York, 1975.

57. Colthup, N. B., and M. K. Orloff, *Spectrochim. Acta*, **27A**, 1299 (1971).

58. Cook, B. W., and K. Jones, *A Programmed Introduction to Infrared Spectroscopy*, Heyden, London, 1972.

59. Cook, D., in G. Olah, Ed. *Friedel-Crafts and Related Reactions*, Vol. I, Interscience, New York, 1963, p. 767.

60. Corish, P. J., *Anal. Chem.*, **31**, 1298 (1959).

61. Cotton, F. A., *Chemical Applications of Group Theory*, Interscience, New York, 1963.

62. Coulson, C. A., *Spectrochim. Acta*, **14**, 161 (1959).

63. Craver, C. D., *Desk Book of Infrared Spectra*, Coblentz Society, POB 9952, Kirkwood, Mo., 1977.

64. Craver, Clara D., Surfaces, in E. G. Brame, Jr., and J. G. Grasselli, Eds., *Practical Spectroscopy*, Vol. 1, Part C, *Infrared and Raman Spectroscopy*, Marcel Dekker, New York, 1977, p. 933.

65. Daasch, L. W., *Spectrochim. Acta*, **13**, 257 (1958).

66. Davis, A., and D. Gordon, *J. Appl. Polym. Sci.*, **18**, 1159 (1974).

67. Davis, J. E. D., *J. Molec. Struct.*, **10**, 1 (1971).

68. Davis, M. A., *J. Org. Chem.*, **32**, 1161 (1967).

69. Deady, L. W., P. M. Harrison, and R. D. Topson, *Spectrochim. Acta*, **31A**, 1671 (1975).

70. Deb, K. K., *Spectrosc. Lett.*, **8**, 185 (1975).

71. Decius, J. C., *J. Chem. Phys.*, **22**, 1941 (1954).

72. Decius, J. C., *J. Chem. Phys.*, **22**, 1946 (1954).

73. Dewar, M. J. S., and P. J. Grisdale, *J. Am. Chem. Soc.*, **84**, 3539, 3548 (1962).

74. Dijkstra, G., *Spectrochim. Acta*, **11**, 618 (1957).

75. Dolphin, D., and A. E. Wick, *Tabulation of Infrared Spectral Data*, Wiley, New York, 1977.

76. Douthit, R. C., K. J. Garska, and V. A. Yarborough, *Appl. Spectrosc.*, **17**, 85 (1963).

77. Drisko, R. W., J. B. Crilly, *Mat. Prot. Perform.*, **11**(12), 49 (1972).

78. Duncan, J. L., Force Constant Calculations in Molecules, in R. F. Barrow, D. A. Long, and D. J. Millen, Eds., *Molecular Spectroscopy,* Vol. 3, The Chemical Society, London, 1975, Chapter 2.

79. Edelmann, K., and H. Wyden, *Kaut. Gummi, Kunstst.,* **25,** 353 (1972).

80. Epp, A., *Anal. Chem.,* **29,** 1283 (1957).

81. Evans, J. C., *Spectrochim. Acta,* **17,** 129 (1961).

82. Farmer, V. C., Infrared Spectroscopy in Mineral Chemistry, in A. W. Nicol, Ed., *Physicochem. Methods of Mineral Analysis,* Plenum Press, New York, 1975.

83. Farmer, V. C., and F. Palmieri, in J. E. Gieseking, Ed., *Soil Components,* Vol. 2, Springer, New York, 1975, p. 573.

84. Fateley, W. G., I. Matsubara, and R. E. Witkowski, *Spectrochim. Acta,* **20,** 1461 (1964).

85. Fateley, W. G., N. T. McDevitt, and F. F. Bentley, *Appl. Spectrosc.,* **25,** 155 (1971).

86. Fermi, E., *Z. Physik,* **71,** 250 (1931).

87. Ferraro, J. R., *Low-Frequency Vibrations of Inorganic and Coordination Compounds,* Plenum Press, New York, 1971.

88. Ferraro, J. R., and J. S. Ziomek, *Introductory Group Theory and Its Application to Molecular Structure,* 2nd ed., Plenum Press, New York, 1975.

89. Fieldes, M., R. J. Furkert, and N. Wells, *N. Z. J. Sci.,* **15,** 615 (1972).

90. Fischmeister, I., L. Hellgren, and J. Vincent, *Arch. Dermatol. Res.,* **253,** 63 (1975).

91. Flett, M. St. C., *Spectrochim. Acta,* **18,** 1537 (1962).

92. Francis, S. A., *J. Chem.,* **18,** 861 (1950).

93. Fraser, R. D. B., and W. C. Price, *Nature,* **170,** 490 (1952).

94. Freeman, N. K., Analysis of Blood Lipids by Infrared Spectroscopy, in G. J. Nelson, Ed., *Blood Lipids and Lipoproteins: Quantity, Composition, and Metabolism,* Wiley, New York, 1972, pp. 113–179.

95. Fritzsche, H., *Spectrochim. Acta,* **17,** 352 (1961).

96. Fuson, N., M. L. Josien, and E. M. Shelton, *J. Am. Chem. Soc.,* **76,** 2526 (1954).

97. Gaffey, M. J., *J. Geophys. Res.,* **81,** 905 (1976).

98. Gay, B. W., Jr., P. L. Hanst, J. J. Bufalini, and R. C. Noonan, *Environ. Sci. Technol.,* **10,** 58 (1976).

99. Gedemer, T. J., *Appl. Spectrosc.,* **19,** 141 (1965).

100. Gerhard, S. L., and D. M. Dennison, *Phys. Rev.,* **43,** 197 (1933).

101. Ghosh, S. N., and A. K. Chatterjee, *J. Mat. Sci.,* **9,** 1577 (1974).

102. Goddu, R. F., and D. A. Delker, *Anal. Chem.,* **32,** 140 (1960).

103. Golding, K., *Process Eng.,* **70,** 1 (1974).

104. Goldstein, M., *Spectrosc. Prop. Inorg. Organomet. Comp.,* **5,** 333 (1972).

105. Gordon, R. G., *J. Chem. Phys.*, **43**, 1307 (1965).

106. Gordy, W., *J. Chem. Phys.*, **14**, 305 (1946).

107. Gordy, W., and W. J. O. Thomas, *J. Chem. Phys.*, **24**, 439 (1956).

108. Gramstad, T., *Spectrochim. Acta*, **19**, 497 (1963).

109. Greenwood, N. N., E. J. F. Ross, and B. P. Straughan, *Index of Vibrational Spectra of Inorganic and Organometallic Compounds*, Vol. 1, Butterworth, London, 1972, and succeeding volumes.

110. Grieble, D. L., P. R. Griffiths, and T. Hirschfeld, *Anal. Chem.*, **50**, 415 (1978).

111. Hadži, D., *Chimia*, **26**, 7 (1972).

112. Hallam. H. E., Solvent Effects on Infrared Group Frequencies, in M. J. Wells, Ed., *Spectroscopy*, Institute of Petroleum, London, 1962, pp. 245–264.

113. Hallam, H. E., and T. C. Ray, *Nature*, **189**, 915 (1961).

114. Hampton, R. R., *Rub. Chem. Technol.*, **45**, 546 (1972).

115. Haslam, J., and H. A. Willis, *Identification and Analysis of Plastics*, Van Nostrand, Princeton, N.J., 1965.

116. Heicklen, J., *Spectrochim. Acta*, **17**, 82 (1961).

117. Helmen, T., *Farbe Lack*, **80**, 715 (1974).

118. Henniker, J. C., *Infrared Spectrometry of Industrial Polymers*, Academic Press, New York, 1967.

119. Herzberg, G., *Molecular Spectra and Molecular Structure. II. Infrared and Raman Spectra of Polyatomic Molecules*, Van Nostrand, New York, 1945.

120. Higuchi, S., S. Tanaka, and H. Kamada, *Spectrochim. Acta*, **28A**, 1721 (1972).

121. Higuchi, S., H. Tsuyama, S. Tanaka, and H. Kamada, *Spectrochim. Acta*, **31A**, 1011 (1975).

122. Hill, A. G., and C. Curran, *J. Phys. Chem.*, **64**, 1519 (1960).

123. Hill, R. R., and D. A. E. Rendell, *The Interpretation of Infrared Spectra*, Heyden, London, 1975.

124. Hinkley, E. D., *Environ. Sci. Technol.*, **11**, 564 (1977).

125. Hirschfeld, T., *Anal. Chem.*, **48**, 721 (1976).

126. Hirschfeld, T., and K. Kizer, *Appl. Spectrosc.*, **29**, 345 (1975).

127. Holland-Moritz, K., and H. W. Siesler, *Appl. Spectrosc. Rev.*, **11**, 1 (1976).

128. Hollas, J. M., *Spectrochim. Acta*, **22**, 81 (1966).

129. Hughes, R. H., R. J. Martin, and N. D. Coggeshall, *J. Chem. Phys.*, **24**, 489 (1956).

130. Hummel, D. O., *Identification and Analysis of Surface-Active Agents*, Wiley-Interscience, New York, 1962.

131. Hummel, D. O., and F. K. Scholl, *Infrared Analysis of Polymers, Resins, and Additives: an Atlas*, Vol. 1, *Plastics, Elastomers, Fibers, and Resins,*

Part 1: Text; Part 2, Spectra, Tables, Index, 1969; Vol. 2, *Additives and Processing Aids,* 1973, Wiley-Interscience, New York.

132. Hummel, D. O., H. Siesler, E. Zoschke, I. Vierling, U. Morlock, and T. Stadtlaender, *Melliand Textilber. Int.,* **54,** 1340 (1973).

133. Hutchinson, B., D. Eversdyk, and S. Olbricht, *Spectrochim. Acta,* **30A,** 1605 (1974).

134. Ingold, C. K., *Structure and Mechanism in Organic Chemistry,* Cornell U. Pr., Ithaca, New York, 1953.

135. Ivin, K. J., *Structural Studies of Macromolecules by Spectroscopic Methods,* Wiley, New York, 1976.

136. Jenkins, J. W., and K. O. Kellenbach, *Anal. Chem.,* **31,** 1056 (1959).

137. Joesten, M. D., and L. J. Schaad, *Hydrogen Bonding,* Marcel Dekker, New York, 1974.

138. Jonas, K., K. Solymar, and M. Orban, *Acta Chim. Acad. Sci. Hung.,* **81,** 443 (1974).

139. Jones, D. W., *Introduction to the Spectroscopy of Biological Polymers,* Academic Press, New York, 1976.

140. Jones, L. A., J. C. Holmes, and R. B. Seligman, *Anal. Chem.,* **28,** 191 (1956).

141. Jones, L. H., *Inorganic Vibrational Spectroscopy,* Vol. 1, Marcel Dekker, New York, 1971.

142. Jones, R. N., *Infrared Spectra of Organic Compounds: Summary Charts of Principal Group Frequencies,* National Research Council of Canada, Ottawa, 1959.

143. Jones, R. N., D. Escolar, J. P. Hawranek, P. Neelakantan, and R. P. Young, *J. Molec. Struct.,* **19,** 21 (1973).

144. Jones, R. N., W. F. Forbes, and W. A. Mueller, *Can. J. Chem.,* **35,** 504 (1957).

145. Jones, R. N., R. Venkataraghavan, and J. W. Hopkins, *Spectrochim. Acta,* **23A,** 925 (1967).

146. Jones, W. J., and N. Sheppard, Rotation in Solution, in M. J. Wells, Ed., *Spectroscopy,* The Institute of Petroleum, London, 1962, pp. 181–204.

147. Kagarise, R. E., *J. Am. Chem. Soc.,* **77,** 1377 (1955).

148. Kagarise, R. E., and K. B. Whetsel, *Spectrochim. Acta,* **18,** 341 (1962).

149. Kappelmeier, C. P. A., *Chemical Analysis of Resin-Based Coating Materials,* Wiley-Interscience, New York, 1959.

150. Karger, B. L., L. R. Snyder, and C. Horvath, *An Introduction to Separation Science,* Wiley, New York, 1973.

151. Karr, C., Jr., *Infrared and Raman Spectroscopy of Lunar and Terrestrial Minerals,* Academic Press, New York, 1975.

152. Katritzky, A. R., and R. D. Topsom, Linear Free Energy Relationships and

Optical Spectroscopy, in N. B. Chapman and J. Shorter, Eds., *Advances in Free Energy Relationships,* Plenum Press, London, 1972.

153. Kawahara, F. K., J. F. Santner, and E. C. Julian, *Anal. Chem.,* **46,** 266 (1974).

154. Kaye, W., Near Infrared Spectroscopy, in G. L. Clark, Ed., *Encyclopedia of Spectroscopy,* Reinhold, New York, 1960, pp. 494–505.

155. Kirkland, J. J., *Anal. Chem.,* **29,** 1127 (1957).

156. Kirkwood, J. G., in W. West and R. T. Edwards, *J. Chem. Phys.,* **5,** 14 (1937).

157. Koenig, J. L., *Appl. Spectrosc.,* **29,** 293 (1975).

158. Koenig, J. L., L. D'Esposito, and M. K. Antoon, *Appl. Spectrosc.,* **31,** 292 (1977).

159. Kollman, P. A., and L. C. Allen, *Chem. Rev.,* **72,** 283 (1972).

160. Krueger, P. J., and H. W. Thompson, *Proc. Roy. Soc. (London),* **A243,** 143 (1957); **A250,** 22 (1959).

161. Krimm, S., *Fortschr. Hochpolym.-Forsch.,* **2,** 51 (1960).

162. Krimm, S., Polymers: Theory and Interpretation of Spectra, in G. L. Clark, Ed., *Encyclopedia of Spectroscopy,* Reinhold, New York, 1960, pp. 535–544.

163. Krimm, S., Infrared Studies of Chain Organization in Crystalline Polyethylene, in E. B. Mano, Ed., *Proceedings of International Symposium on Macromolecules,* 1974, Elsevier, Amsterdam, 1975, p. 107.

164. Kuhn, L. P., and R. E. Bowman, *Spectrochim. Acta,* **17,** 650 (1961).

165. Langseth, A., and R. C. Lord, *Kgl. Danske Videnskab. Selskab, Mat-fys. Medd.,* **16,** 6 (1948).

166. Larson, H. P., and U. Fink, *Appl. Spectrosc.,* **31,** 386 (1977).

167. Larson, H. P., and U. Fink, *Astrophys. J. Lett.,* **171,** L91 (1971).

168. Launer, P. J., *Perkin-Elmer Instrum. News,* **13,** (3), 10 (1962).

169. Layton, E. M., Jr., R. D. Kross, and V. A. Fassel, *J. Chem. Phys.,* **25,** 135 (1956).

170. LeBlanc, R. B., and T. T. Trimmer, *Text. Chem. Color,* **5,** 279 (1973).

171. Leffler, J. E., and E. Grunwald, *Rates and Equilibria of Organic Reactions,* Wiley, New York, 1963.

172. Levi, L., and C. E. Hubley, *Anal. Chem.,* **28,** 1591 (1956).

173. Levitt, L. S., and H. F. Widing, The Alkyl Inductive Effect, in R. W. Taft, Ed., *Progress in Physical Organic Chemistry,* Wiley, New York, Vol. 12, 1976, p. 119.

174. Liddel, U., III, and E. D. Becker, *Spectrochim. Acta,* **10,** 70 (1957).

175. Linton, H. R., and E. R. Nixon, *J. Chem. Phys.,* **28,** 990 (1958).

176. Linton, H. R., and E. R. Nixon, *Spectrochim. Acta,* **10,** 299 (1958).

177. Lippert, E., and R. Mecke, *Z. Elektrochem.,* **55,** 366 (1951).

178. Lisitsa, M. P., and Yu. P. Tsyaschenko, *Opt. Spektroskopiya,* **9,** 438 (1960); *Opt. Spectrosc.,* **9,** 229 (1960) (transl.).

179. Lord, R. C., and R. E. Merrifield, *J. Chem. Phys.,* **21,** 166 (1953).

180. Lord, R. C., and R. W. Walker, *J. Am. Chem. Soc.,* **76,** 2518 (1954).

181. Lunn, W. H., *Spectrochim. Acta,* **16,** 1088 (1960).

182. Luongo, J. P., *Appl. Spectrosc.,* **25,** 76 (1971).

183. Lynch, P. F., S. Tang, and C. W. Brown, *Anal. Chem.,* **47,** 1696 (1975).

184. Maguire, M. M., and R. West, *Spectrochim. Acta,* **17,** 369 (1961).

185. Mantz, A. W., and H. K. Morita, *Appl. Spectrosc.,* **30,** 587 (1976).

186. Maslowsky, E., Jr., *Vibrational Spectra of Organometallic Compounds,* Wiley, New York, 1976.

187. McCall, E. R., S. H. Miles, V. W. Tripp, and R. T. O'Connor, *Appl. Spectrosc.,* **18,** 81 (1964).

188. McCarthy, D. E., *Appl. Spectrosc.,* **22,** 66 (1968).

189. McGraw-Hill, Inc., *Modern Plastics Encyclopedia,* McGraw-Hill, New York, Annual.

190. McKean, D. C., *Spectrochim. Acta,* **29A,** 1559 (1973).

191. McKean, D. C., *Spectrochim. Acta,* **31A,** 1167 (1975).

192. McKean, D. C., S. Biedermann, and H. Bürger, *Spectrochim. Acta,* **30A,** 845 (1974).

193. McKean, D. C., O. Saur, J. Travert, and J. C. Lavalley, *Spectrochim. Acta,* **31A,** 1713 (1975).

194. Meiklejohn, R. A., R. J. Meyer, S. M. Aronovic, H. A. Schuette, and V. W. Meloch, *Anal. Chem.,* **29,** 329 (1957).

195. Morris, N. M., E. R. McCall, and V. W. Tripp, *Text, Chem. Color,* **4,** 283 (1972).

196. Münch, W., U. Fringeli, and Hs. H. Günthard, *Spectrochim. Acta,* **33A,** 95 (1977).

197. Murthy, A. S. N., and C. N. R. Rao, *Appl. Spectrosc. Rev.,* **2,** 69 (1968).

198. Myers, R. R., and J. S. Long, *Treatise on Coatings: Film-Forming Compositions,* Part I, Marcel Dekker, New York, 1967.

199. Nakanishi, K., and P. H. Solomon, *Infrared Absorption Spectroscopy,* 2nd ed., Holden-Day, San Francisco, 1977.

200. Nakomoto, K., *Infrared Spectra of Inorganic and Coordination Compounds,* 2nd ed., Wiley-Interscience, New York, 1970.

201. Nasser, M. I., *Appl. Spectrosc.,* **28,** 545 (1974).

202. Nyquist, R. A., and W. J. Potts, Jr., Vibrational Spectra of Phosphorus Compounds, in M. Halman, Ed., *Analytical Chemistry of Phosphorus Compounds,* Interscience, New York, 1972.

203. O'Connor, R. T., *High Polym.,* **5,** 51 (1971).

204. O'Connor, R. T., E. R. McCall, N. M. Morris, and V. W. Tripp, *U.S. Agric. Res. Serv., South. Reg. (Rep.),* **1974,** ARS-S-47 (1974).

205. Overend, J., and J. R. Scherer, *Spectrochim. Acta,* **16,** 773 (1960).

206. Parker, F. S., *Applications of Infrared Spectroscopy in Biochemistry, Biology, and Medicine,* Plenum Press, New York, 1971.

207. Parker, F. S., *Appl. Spectrosc.,* **29,** 129 (1975).

208. Pauling, L., *Nature of the Chemical Bond,* 3rd ed., Cornell U. P., Ithaca, N. Y., 1960.

209. Phillips, J. P., and J. F. Deye, *Anal. Chim. Acta,* **17,** 231 (1957).

210. Pimentel, G. C., and A. L. McClellan, *Annu. Rev. Phys. Chem.,* **22,** 347 (1971).

211. Pimentel, G. C., and A. L. McClellan, *The Hydrogen Bond,* Freeman, San Francisco, 1960.

212. Pinchas, S., and I. Laulicht, *Infrared Spectra of Labeled Compounds,* Academic Press, New York, 1971.

213. Pitts, J. N., B. J. Finlayson-Pitts, and A. M. Winer, *Environ. Sci. Technol.,* **11,** 568 (1977).

214. Polčin, J., *Appl. Spectrosc.,* **28,** 588 (1974).

215. Potts, W. J., Jr., *Chemical Infrared Spectroscopy,* Vol. 1, *Techniques,* Wiley, New York, 1963.

216. Potts, W. J., Jr., The Use of Infrared Spectroscopy in Characterization of Polymer Structure, in *Symposium on Plastics Testing and Standardization,* ASTM Special Technical Publication No. 247, 1958, p. 225.

217. Potts, W. J., Jr., and R. A. Nyquist, *Spectrochim. Acta,* **15,** 679 (1959).

218. Powell, D. L., and R. West, *Spectrochim. Acta,* **20,** 983 (1964).

219. Pritchard, H. O., and H. A. Skinner, *Chem. Rev.,* **55,** 745 (1955).

220. Pros, A. W., and F. van Zeggeren, *Spectrochim. Acta,* **16,** 563 (1960).

221. Pullin, A. D. E., *Proc. Roy. Soc. (London),* **A255,** 39 (1960).

222. Puttnam, N. A., *J. Soc. Cosmet. Chem.,* **23,** 209 (1972).

223. Rakov, A. V., *Opt. Spektroskopiya,* **13,** 369 (1962); *Opt. Spectrosc.,* **13,** 203 (1962) (transl.).

224. Ramsay, D. A., *J. Am. Chem. Soc.,* **74,** 72 (1952).

225. Rezinkova, E. B., V. T. Tulin, and V. M. Tatevskii, *Opt. Spektroskopiya,* **13,** 364 (1962); *Opt. Spectrosc.,* **13,** 200 (1962) (transl.).

226. Ridgway, S., H. P. Larson, and U. Fink, The Infrared Spectrum of Jupiter, in T. Gehrels, Ed., *Jupiter,* Arizona U. P., Tucson, 1976.

227. Rosen, M. J., and H. A. Goldsmith, *Systematic Analysis of Surface-Active Agents,* 2nd ed., Wiley-Interscience, New York, 1972.

228. Sandorfy, C., *Can. J. Spectrosc.,* **17,** 24 (1972).

229. Scherer, J. R., and J. Overend, *J. Chem. Phys.,* **32,** 1720 (1960).

230. Scherer, J. R., and J. Overend, *J. Chem. Phys.,* **33,** 1681 (1960).

231. Schmid, P., *Physiol. Chem. Phys.*, **7**, 335, 349, 357 (1975).

232. Schwarzmann, E., *Naturwissenschaft.*, **49**, 103 (1962).

233. Seshadri, K. S., and R. N. Jones, *Spectrochim. Acta*, **19**, 1013 (1963).

234. Sheppard, N., Rotational Isomerism about C—C Bonds in Saturated Molecules Studied by Vibrational Spectroscopy, in H. W. Thompson, Ed., *Advances in Spectroscopy*, Vol. I, Interscience, New York, 1959, pp. 288–353.

235. Shimanouchi, T., *Tables of Molecular Vibrational Frequencies, Consolidated Volume I*, U.S. Government Printing Office, S. D. Cat. No. C13.48:39, 1972.

236. Shimizu, H., *Bull. Chem. Soc. Jap.*, **39**, 2385 (1966).

237. Shipman, J. J., V. L. Folt, and S. Krimm, *Spectrochim. Acta*, **18**, 1603 (1962).

238. Shuster, P., G. Zundel, and C. Sandorfy, *The Hydrogen Bond: Recent Developments in Theory and Experiments*, Vol. 2, *Structure and Spectroscopy*. Elsevier North-Holland, Amsterdam, 1976.

239. Silverstein, R. M., G. C. Bassler, and T. C. Morrill, *Spectrometric Identification of Organic Compounds*, 3rd ed., Wiley, New York, 1974.

240. Slowinski, E. J., Jr., and G. C. Claver, *J. Opt. Soc. Am.*, **45**, 396 (1955).

241. Smith, A. L., *Am. Lab.*, **2**, (10), 27 (1970).

242. Smith, A. L., Infrared Spectroscopy, in J. W. Robinson, Ed., *Handbook of Spectroscopy*, Vol. II, CRC Press, Cleveland, 1974.

243. Spialter, L., and M. Ballester, *Anal. Chem.*, **34**, 1183 (1962).

244. Steele, D., *Theory of Vibrational Spectroscopy*, Saunders, Philadelphia, 1971.

245. Stock, Leon M., *J. Chem. Educ.*, **49**, 400 (1972).

246. Strizhevskii, V. L., *Opt. Spektroskopiya*, **8**, 165 (1960); *Opt. Spectrosc.*, **8**, 86 (1960) (transl.).

247. Susz, B. P., and A. Lachavanne, *Helv. Chim. Acta* **41**, 634 (1958).

248. Swain, C. G., and E. C. Lupton, Jr., *J. Am. Chem. Soc.*, **90**, 4328 (1968).

249. Sward, G. G., and H. A. Gardner, *Paint Testing Manual: Physical and Chemical Examination of Paints, Varnishes, Lacquers, and Colors*, 13th ed., American Society for Testing and Materials, Philadelphia, 1972.

250. Tabb, D. L., J. J. Sevcik, and J. L. Koenig, *J. Polym. Sci., Polym. Phys. Ed.*, **13**, 815 (1975).

251. Taft, R. W., Jr., Separation of Polar, Steric, and Resonance Effects in Reactivity, in M. S. Newman, Ed., *Steric Effects in Organic Chemistry*, Wiley, New York, 1956, pp. 556–675.

252. Thomas, G. J., Jr. and Y. Kyogoku, Biological Science, in E. G. Brame, Jr., and J. G. Grasselli, Eds., *Practical Spectroscopy*, Vol. 1, *Infrared and Raman Spectroscopy*, Part C, Marcel Dekker, New York, 1977, p. 717.

253. Thomas, L. C., *Interpretation of the Infrared Spectra of Organophosphorus Compounds*, Heyden, London, 1974.

254. Thompson, B., *Hazardous Gases and Vapors: Infrared Spectra and Physical Constants*, Beckman Instruments, Fullerton, Ca., Technical Report No. 595, August 1974.

255. Thompson, H. W., and D. A. Jameson, *Spectrochim. Acta*, **13**, 236 (1958).

256. Thompson, H. W., and D. J. Jewell, *Spectrochim. Acta*, **13**, 254 (1958).

257. Topsom, R. D., The Nature and Analysis of Substituent Electronic Effects, in R. W. Taft, Ed., *Progress in Physical Organic Chemistry*, Vol. 12, Wiley, New York, 1976.

258. Tosi, C., and F. Ciampelli, *Fortschr. Hochpolym.-Forsch.*, **12**, 87 (1973).

259. Tweed, F. T., R. Cameron, J. S. Deak, and P. G. Rogers, *Forens. Sci.*, **4**, 211 (1974).

260. Van der Maas, J. H., and E. T. G. Lutz, *Spectrochim. Acta*, **30A**, 2005 (1974).

261. Van der Marel, H. W., and H. Beutelspacher, *Atlas of Infrared Spectroscopy of Clay Minerals and Their Admixtures*, Elsevier, New York, 1976.

262. Venkateswarlu, K., and S. Jagathezen, *Opt. Spektroskopiya*, **13**, 775 (1962); *Opt. Spectrosc.*, **13**, 439 (1962) (transl.).

263. Washburn, W. H., *Appl. Spectrosc.*, **11**, 46 (1957).

264. Weiner, M., G. Vogel, and R. West, *Inorg. Chem.*, **1**, 654 (1962).

265. West, R., and W. Glaze, *J. Am. Chem. Soc.*, **83**, 3580 (1961).

266. Whetsel, K. B., *Appl. Spectrosc. Rev.*, **2**, 1 (1968).

267. Whetsel, K. B., and R. E. Kagarise, *Spectrochim. Acta*, **18**, 315, 329 (1962).

268. Wiberley, S. E., S. C. Bunce, and W. H. Bauer, *Anal. Chem.*, **32**, 217 (1960).

269. Wilkins, P. E., *Air Qual. Instrum.*, **2**, 246 (1974).

270. Wilks, P. A., Jr., *Environ. Sci. Technol.*, **10**, 1204 (1976).

271. Wilmshurst, J. K., *J. Chem. Phys.*, **28**, 733 (1958).

272. Wilson, E. B., Jr., J. C. Decius, and P. C. Cross, *Molecular Vibrations*, McGraw-Hill, New York, 1955.

273. Woodward, L. A., *Introduction to the Theory of Molecular Vibrations and Vibrational Spectroscopy*, Oxford U. P., London, 1972.

274. Young, C. W., R. B. Duvall, and N. Wright, *Anal. Chem.*, **23**, 709 (1951).

275. Zbinden, R., *Infrared Spectroscopy of High Polymers*, Academic Press, New York, 1964.

276. Zenker, W., *Anal. Chem.*, **44**, 1235 (1972).

QUANTITATIVE APPLICATIONS

USE OF IR IN QUANTITATIVE ANALYSIS

One of the most significant advantages of an IR spectrum is its uniqueness. This specificity is a great asset also in quantitative analysis; often individual bands can be found for each component of a mixture that are subject to little or no interference from the other components. If interferences do occur, they may usually be corrected for by standard techniques.

A second advantage is that semiquantitative data may be obtained from the qualitative spectrum of a sample. This fact is overlooked by many occasional users of analytical techniques, who may redundantly obtain both IR spectra and gas chromatograms in an attempt to get qualitative and semiquantitative information.

Infrared can also be applied quantitatively to an extremely wide variety of samples; it is not restricted to volatile, crystalline, or soluble materials. In many cases, extensive sample preparation is not necessary. It has been used for widely different problems such as determination of unsaturation in polyethylene [93], quartz particles in air [107], D_2O in H_2O [36], CO_2 in wine [24]; CO in blood and heart muscle [76], and fat, protein, and sugar in milk [1].

Disadvantages to the IR approach include occasional serious interferences; transparency of homonuclear molecules, time required for sample preparation, need for prior standards, nontransferability of calibration data between different spectrometers, and (in some cases) lack of sensitivity. Gas chromatography, because of its convenience, accuracy, speed, and sensitivity, has replaced IR spectroscopy as a quantitative tool for the analysis of many gases and volatile liquids. Nuclear magnetic resonance spectroscopy offers some unique attractions for certain types of liquid or soluble materials, as does mass spectroscopy. Which of these techniques is chosen in a particular case depends on the system to be analyzed, the tools available, the accuracy required, and the amount of sample preparation needed.

When planning a quantitative IR analysis, one should first ask how precise the determination must be. The usual answer to this question is "as accurate as possible"—a completely ambiguous statement. A little probing will usually reveal realistic tolerances for error in the determination, and when this is done one can proceed with the development. This point is emphasized because the time, complexity, and cost of doing an analysis varies roughly in proportion to the desired accuracy, and it is wasteful to report results that are more accurate than needed. Semiquantitative analyses, with answers accurate to about ±10% relative, are easy to carry out and require only single standards on the pure components (or, in special cases, no standards at all). Quantitative analyses with accuracies on the order of ±1% demand a number of precautions, usually including careful preparation of synthetic standards. Another magnitude of accuracy is attainable under favorable conditions but requires studious attention to detail and utmost care in manipulation.

In the following discussion we briefly consider standard techniques common to all infrared analyses and then explore more fully certain specialized areas such as high-accuracy determinations, trace analyses, and functional-group analysis.

This introduction would not be complete without recognizing the frequently repeated truism that an analysis can be no better than the sample. Sampling of large quantities of material is beyond the scope of this chapter and often beyond the control of the analyst, but proper mixing and selection from the IR sample is the responsibility of the spectroscopist. Liquids, gases, and powders should not be assumed to be homogeneous; thorough mixing is important. When sufficient material is available, at least a gram should be weighed into solution in the IR solvent, even though only a small fraction of this amount is actually used. Solids and polymers present a more difficult problem and considerable ingenuity often must be used to obtain a representative sample. A bibliography that covers sampling of a variety of materials is given by Walton and Hoffman [112].

LAWS OF ABSORPTION

The physical laws governing the absorption of radiation are discussed in detail elsewhere [79]; therefore, the treatment given here is confined to definitions of terms.

Bouguer's Law

The law stated originally by Bouguer [18] and later independently by Lambert relates the absorption of a material to its thickness. It states that each layer of equal thickness absorbs a constant fraction of the radiant energy passing through it. It is an exact law for a homogeneous system. Stated mathematically

$$-\frac{dI}{db} = a_t I \tag{6.1}$$

where I is the intensity or energy of the radiation and b is the unit of thickness. The factor a is a proportionality constant related to the absorptive power of the sample. In its integrated form the relationship is

$$\ln \frac{I_0}{I} = a_t b \tag{6.2}$$

where I_0 is the intensity of the incident and I is the intensity of the transmitted radiation.

Beer's Law

The relationship between monochromatic light absorption of a species and its concentration in a homogeneous medium was formulated by Beer [14]. It states that the absorption of a species is proportional to its concentration, c:

$$-\frac{dI}{dc} = a_c I \tag{6.3}$$

or

$$\ln \frac{I_0}{I} = a_c c \tag{6.4}$$

Bouguer–Beer Law

The combined relationship is written

$$-\log \frac{I}{I_0} = \log \frac{I_0}{I} = abcc' = A \tag{6.5}$$

We have, for later convenience, written the concentration term as the product of two factors: c is the fraction of the desired constituent in the sample, and c' is the concentration of sample in the analytical solution. The term $\log I_0/I$ is commonly known as the *absorbance*, A (formerly the optical density), and the constant a is the *absorptivity*. The *transmittance* is I/I_0 and the *percent transmittance* is $100 I/I_0$. Logarithms to the base 10 are ordinarily used, with the factor 2.3 being incorporated into the constant a.

Equation (6.5) is fundamental to all spectrometric quantitative analyses. It says that a straight-line relationship should be obtained from a plot of absorbance against concentration or thickness of a given species in a homogeneous mixture or solution. It further predicts that, at the same wavelength, absorbances of bands from different species will be additive (this is the law of additivity).

Deviations from Absorption Laws

Deviations from equation (6.5) occur frequently, but *they need not affect quantitative accuracy* if they are recognized. The causes for such discrepancies may be external (spectrometer and cell imperfections) or internal (systemic effects).

Probably the largest single factor causing deviations from the Bouguer–Beer law is the finite slit width necessarily used on all spectrometers. As we saw earlier (Chapter 2), the combined effect of the slit width and the particular slit function of a spectrometer is that a band of neighboring frequencies, rather than monochromatic radiation, reaches the detector. Furthermore, the spread of frequencies becomes wider as one opens the slits to reduce the noise level. Since the law [equation (6.5)] holds only for monochromatic radiation, discrepancies are expected unless the slit width is much less than the band width. The ratio of the true to the apparent maximum absorbance for various absorbances and slit widths has been calculated and tabulated by Ramsay [90] for Lorentzian bands. At an absorbance of 1, for example, and with a slit:band width ratio $S:\Delta\nu_{1/2} = 0.2$, A (true):A (observed) $= 1.03$; for $S:\Delta\nu_{1/2} = 0.5$, A (true):A (observed) $= 1.24$. Clearly, then, even small *changes* in slit width (or slit function) of a spectrometer used in quantitative analysis will affect the slope and linearity of the absorbance–concentration curves.

Other instrumental factors important in producing apparent deviations from the Bouguer–Beer law include sample reflection and scattering losses, beam-convergence effects, and errors in setting the spectrometer zero. They are discussed more fully later.

Systemic errors arise from the failure of the absorbing species to retain its identity in a homogeneous solution. Common manifestations of this effect include intermolecular association, especially hydrogen bonding, and micelle formation. Quantitative analysis of associated species is discussed later. In materials run as solids, insufficiently ground samples show large Beer's law deviations, since sharpness of the bands depends strongly on the fineness of the powder (see pp. 251–252).

THE PRACTICE OF QUANTITATIVE ANALYSIS

During 1957–1960, 150 different quantitative analytical methods were published by the journal *Analytical Chemistry,* under sponsorship of the Coblentz Society. Subsequent methods have been carried by *Applied Spectroscopy.* Anyone contemplating development of a quantitative IR analysis might first consult an index to the methods found in *Applied Spectroscopy* [103].

All quantitative IR analyses are carried out by comparing, directly or indirectly, the absorbance of the unknown at a given wavelength (often the peak of a strong absorption band) with the absorbance of the same material in a standard of known concentration. Peak absorbance, since it is easily measured and is directly relatable to concentration, is the most useful parameter to use in calculations. Absorbance measurements on the sides of bands should be avoided, since very small errors in reproducing the wavelength setting produce large absorbance variations. Any band, strong or weak, can be used for analysis, provided its absorbance is brought into an optimum range by proper selection of solution concentration and cell thickness. It is advantageous to choose bands having minimum interference by other absorptions in the spectrum.

As we noted in equation (6.5), absorbance is proportional to the logarithm of the transmittance—an awkward function to use in routine measurements. In reading absorbance from a chart, one might wish for a record linear in absorbance much the same as is given by an UV spectrometer because such a presentation would simplify data taking. Furthermore, since absorbances are additive, the total absorbance at a given wavelength is simply the sum of the absorbances of the individual species. There are several reasons why IR spectrometers usually present data in terms of percent transmittance, however. First, the electrical signal that is amplified to produce the spectral record is linear with percent transmittance. Although it is perfectly feasible to modify the output with a mechanical or electrical cam so that the pen record reads absorbance directly, this is not usually done because of the greatly increased noise level that would result at high absorbance in the energy-limited IR region (Fig. 6.1). Another reason for using the linear percent transmittance ordinate is that weaker bands are accentuated, strong bands attenuated, and even the strongest absorptions do not go off scale.

Absorbances can be measured by using an ordinary millimeter scale and taking the logarithm of the ratio I_0/I (Fig. 6.2). An easier method is to use a transparent ruler calibrated in absorbance. The point $A = \infty$ is superimposed on the true 0% transmittance point of the chart and the

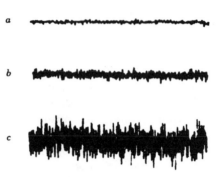

Fig. 6.1 Equivalent noise levels at absorbances of zero (*a*), one (*b*), and two (*c*) on linear absorbance scale.

values at I_0 and I are read from the scale. The difference between these numbers is the absorbance of the band. The same principle is used with spectrometer charts printed with an absorbance ordinate (Fig. 6.2).

Integrated absorbance measurements, in which the total area of the band is measured, are seldom used for quantitative analysis because peak absorbance measurements can be made much more accurately and reproducibly, at least on the same spectrometer. Also, true integrated absorbance measurements require either linear absorbance ordinate re-

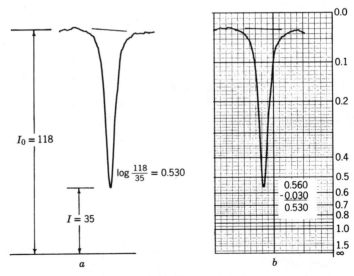

Fig. 6.2 Absorbance measurement using millimeter scale (*a*) and absorbance read directly from chart paper (*b*).

cording or tedious hand replotting of percent transmittance values on an absorbance scale.

For quantitative analysis, the 0% transmittance point should be located as accurately as possible before any transmittance measurements are made. It should be set to correspond to the $A = \infty$ (or percent transmittance = 0) line of the chart paper. (The percent transmittance = 100 point, however, is best set at 90–95% on the transmittance scale.) Accurate zero settings are difficult to achieve with double-beam optical null spectrometers. As the sample beam is shuttered, the optical attenuator moves toward the 0% transmittance position and consequently little or no energy is available to activate the servo system. The zero is then "dead" and consequently cannot be located with precision. Ratio-recording spectrometers do not, of course, present this problem in setting the zero point. The first step in the zero setting is to assure that the pen drift is negligible; in other words, when both beams are completely blocked with the pen at midscale, no pen motion, other than normal noise, occurs. Both shutters are then opened and the sample shutter is closed *very slowly* as the pen approaches zero to avoid overshooting because of inertia in the pen servo system. The shutter used might well be a material that is opaque at that wavelength but transparent to higher frequencies, as described earlier (pp. 38–39). Alternatively, a heavy layer of sample (10–100 times thicker) or another material with strong absorption near that wavelength (Table 2.1) may be scanned to record a band of virtually infinite absorbance. Such bands should be scanned at very slow speeds to minimize error from pen overshoot. An alternative method that reputedly gives greater accuracy for normal quantitative analysis is described later.

Semiquantitative Analysis

Among the advantages of using solution sampling and keeping the spectrometer well-tuned and in calibration is the fact that spectra obtained under these conditions can be used directly for semiquantitative analysis. A spectrum of the pure material is used as reference, and impure materials or components of mixtures can usually be analyzed with an accuracy of ±10% relative or better.

An absorbance measurement, as we have seen, involves measuring not only the transmittance (I) at the band minimum, but also a determination of the background intensity (I_0). This point may be difficult to determine accurately, but usually a satisfactory "baseline" can be constructed [51, 88, 121]. This method of determining I_0 is at best approximate, but it is easy to carry out, reproducible, and of wide applicability. The spectrum is scanned over the analytical band and a straight "baseline" is drawn between points of no absorption on either side of the absorption maximum (Fig. 6.3a). The real utility of the method is seen, however, when the band to be measured falls on the side of another absorption (Fig. 6.3b). In this case the baseline is drawn tangent to the spectrometer tracing as shown.

Usually the most difficult part of using the baseline method is trying to determine the end points. The criterion is to *draw the baseline as nearly as possible where the pen tracing would go if the band were not present.* This approximation may at times be very poor (as for band *b* in Fig. 6.3c). When a determination is set up using a series of standards and a calibration curve is prepared, the actual baseline points do not matter much, provided they are used consistently and provided that no

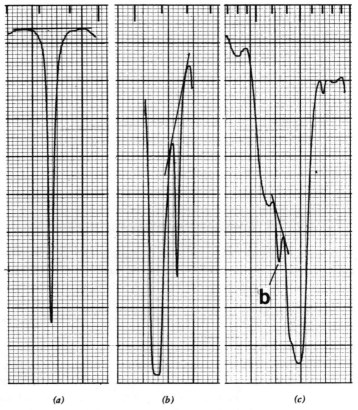

(a)   (b)   (c)

Fig. 6.3 (a) Baseline construction for isolated band; (b) baseline drawn against interfering band; (c) baseline that poorly approximates background.

new interfering bands appear. Any systematic error is automatically compensated for by the calibration curve. A discussion of baseline constructions is given by Potts [88, Chapter 6].

If occasional semiquantitative analysis is being carried out using spectra on pure materials as standards, quite large errors may result from baseline ambiguities. In such cases one may succumb to the temptation to sketch in a background that presumably approximates the true background. This process can be made slightly less subjective by assuming a Lorentzian or at least a symmetrical shape for the overlapping bands and by using tracing paper to draw a series of bands that, when added together, give the observed contour. Such estimations, however, may be no more accurate than the baseline method and are certainly less reproducible.

Normal-precision Quantitative Analysis

We stated that one of the principal causes of deviations from Beer's law is the finite slit width of the IR spectrometer. Nevertheless, optimum accuracy in quantitative analysis is obtained by opening the slits by a factor of at least 2, and preferably 3. The reason, succinctly put, is that we can cope with Beer's law deviations, but not with spectrometer noise. The best remedy then, is to increase the signal:noise ratio by opening the slits [recall from equation (2.7) in chapter 2 that $S/N \propto w^2$] and accept the resulting slight deterioration in peak shape [83, 88]. In fact, for a good grating spectrometer the loss of resolution will be quite small, since normal slit widths are less than band widths for most liquids.

A *reproducible* (but not necessarily accurate) zero setting is essential for precise quantitative work. To make this setting with a double-beam optical null spectrometer, Perry [83] recommends using a screen with about 15% transmittance. The screen is carefully cleaned and used in a holder such that it can be reproducibly inserted into the sample beam, well away from the beam focus. The screen is treated as an optical element; that is, it is carefully handled and the optical portion is not touched. Its transmittance is determined approximately; thereafter, this figure is assumed to be exact and the spectrometer zero is adjusted accordingly. Any difference between the true and assumed 0% transmittance point is automatically accounted for in the calibration curve. Once the spectrometer controls have been set for a given set of measurements, they should not be changed until the measurements are finished. The dramatic effects of slit width and zero setting on absorbance error are shown in Fig. 6.4.

The background or I_0 line may be established by using baseline tech-

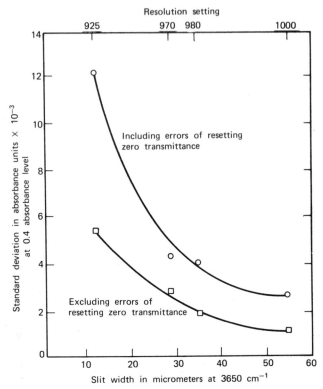

Fig. 6.4 Absorbance error as function of slit width and error in resetting 0% transmittance point [83].

niques discussed previously, by refilling the sample cell with pure solvent (or using a similar solvent-filled cell as reference) or, if possible, by choosing a nearby point in the spectrum where no absorption occurs. Again, any difference between the measured I_0 and the true I_0 will be compensated in the calibration curve.

In the special case of a two-component mixture or in cases where the concentration of one component is constant, one can use the ratio of two band absorbances instead of working curves based on corrected absorbances. Cell thickness and solution concentration need not be measured, since these factors cancel; thus the method is applicable to polymer films and samples run by ATR as well as to liquids and solutions.

Because of the rapid changes in refractive index of a liquid in the vicinity of an absorption band quantitative analysis should be carried out in dilute solution, preferably 2% w/v or less [42]. Samples are weighed into volumetric flasks (see pp. 73–76), using accuracy appropriate to the situation, and the flask is immediately filled to the mark with solvent. The analytical bands selected should preferably be isolated, that is, have no interference from the other components or from the solvent. They need *not* be the strongest bands in the spectrum. For minimum error in the absorbance reading, a cell thickness should be chosen to give analytical band absorbances near 0.43 or 37% transmittance; in practice, a range of 0.2–0.7 (20–60% transmittance) is quite satisfactory. If the absorbance-ratio method is used, the optimum absorbance depends on the relative intensities of the two bands as shown in Fig. 6.5. Cell-window materials should be selected such that their refractive index matches that of the solvent. Cell thickness may be determined by interference fringes or by reference to a readily available pure material that can be run undiluted in a cell that has been calibrated by the fringe method. The absorbance A of a given band will then be proportional to the cell thickness t; that is,

$$\frac{A_1}{A_2} = \frac{t_1}{t_2} \tag{6.6}$$

where subscripts 1 and 2 refer to the two cells. Such measurements should, of course, be made using instrument parameters optimized for quantitative analysis.

In routine quantitative analyses in which the identity of all the components is known, either of two systems may be used to obtain absorbances of the bands to be measured. In the scanning method the spectrometer is set initially to a wavelength at which no absorption occurs and is allowed to scan slowly and completely over the absorption band. If the scanning rate is too fast, the pen will not accurately reproduce the band contour and the absorbance will be too low. Scanning has the advantage of showing unexpected shifts in the absorption maximum or the presence of extraneous absorptions that may affect the measurement.

In the single-point-measurement method, the spectrometer is set to the wavelength of the band maximum and the absorbance is read on the chart or the ordinate scale. One advantage of the point-measurement technique is that by using electrical filtering with a long time constant, one can average out noise fluctuations over a much longer period than is possible when scanning. Furthermore, the overall time required to obtain the measurement may be somewhat less. Obviously the labor involved in setting up a point-measurement method justifies its use only for routinely analyzed samples. Samples run by single-point measurements should be scanned occasionally to detect any unexpected changes in constitution.

Point measurements can be easily made using single-beam spectrom-

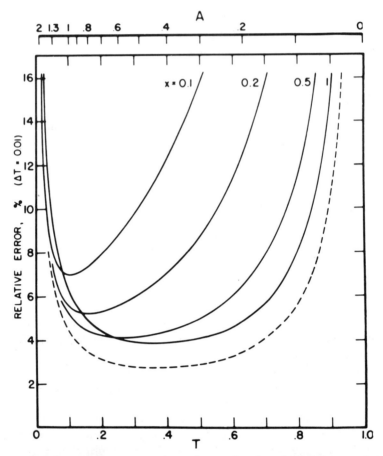

Fig. 6.5 Relative error in concentration ratio x as function of transmittance of stronger peak, T, for $x = 0.1$, 0.2, 0.5, and 1 (for $\Delta T = 0.01$). Dashed line is the Twyman–Lothian curve for relative error in concentration determined from a single peak. Reproduced, with permission, from *Applied Spectroscopy* [77].

eters, and it appears that the accuracy of such measurements is slightly better than is obtained from double-beam spectrometers.

Summary

The procedure suggested in the following list is intended as a guide to development of routine quantitative IR analytical methods. Naturally, no one procedure is universally applicable, and the experience and skill of the spectroscopist must be relied on to adapt or modify the development as necessary. It is also assumed that the analyst has considered various analytical possibilities for solving his problem and has decided that IR is the most advantageous technique. Development of the method might well be based on the following outline:

Define the problem: How many components are present? What are they? Which of them need to be determined? With what accuracy? What are the expected concentration limits for each component? This step will involve running a number of preliminary spectra to obtain a semiquantitative picture of the system.

Choose optimum conditions: Run spectra of each constituent. Select analytical bands showing minimum interference. Choose a combination of cell thickness and concentration(s) such that solvent absorption is negligible and that analytical bands for constituents at their median concentrations fall in the 20–60% transmittance range. Select a signal:noise ratio somewhat better than required to meet accuracy specifications; determine slit width or program (usually 2–3× normal), adjust spectrometer response, and check several scan speeds to find the optimum rate.

Prepare standards: At least four are required for each component. Check the linearity of the absorbance–concentration plot. Prepare mixed standards, if several components are to be determined. Check the law of additivity. Choose the best method of calculating concentrations. Determine interference corrections, if necessary.

Run Samples. Use instrumental conditions and handling techniques identical to those used during standardization. Check accuracy frequently by analysis of standards or known samples. Evaluate results statistically.

High-precision Quantitative Analyses

Precision may be further increased by an extension of the techniques described in the preceding paragraphs. The importance of proper sampling has already been stressed, but it is important to realize that to attain, say, 0.5% accuracy in an analysis, the sample must be representative and homogeneous at least to that degree, even if no margin is allowed for other errors. As greater accuracy is sought, the need for representative sampling becomes progressively greater and, in many instances, the precision of an analysis is limited by lack of sample homogeneity.

In work of highest accuracy, evaporation errors may become appreciable. This is particularly true if the vapor pressure of the solute is high at room temperature, for then the concentration may change rapidly on exposure to air. If samples and standards are handled in the same manner, however, manipulative and evaporation errors should affect both to about the same degree.

A related problem occurs with moisture- or oxygen-sensitive materials, which may require manipulation in a dry box or under inert atmosphere, as appropriate. Special precautions may be necessary with water-reactive species, including the use of well-dried glassware and cells. In any case, such materials should have minimum exposure to the atmosphere. It is also advantageous to use moderately dilute solutions, say, 2%, and to keep all containers as full as possible and tightly closed. Absorbance is sensitive to sample temperature, as discussed in the following section on errors.

We noted in a preceding discussion that opening the spectrometer slit by a factor of w makes possible a decrease in the noise level by a factor of w^2. Further increases in analytical precision can be obtained by the use of difference spectroscopy combined with ordinate (pen) expansion.

Standard-compensated double-beam or difference spectroscopy (sometimes called "differential" spectroscopy) has been practiced in the UV and visible ranges for many years but has not had widespread application in the IR region. The reasons for this are twofold: until recently, instrumental limitations have greatly restricted its potential; and the very small sample thickness used in the IR makes exact compensation difficult. Nevertheless, for analyses of highest accuracy or for certain kinds of trace analysis, the technique of difference spectra is unsurpassed.

The term "difference spectra" is favored here because "differential" implies infinitesimally small differences and is thus an inaccurate description of the situation. Furthermore, it could be confused with derivative spectroscopy, which is concerned with recording the first or second derivative of the absorption curve [28].

An accessory device for IR spectrometry that is particularly useful in conjunction with difference spectroscopy is ordinate or pen expansion. It is available commercially or can be constructed in the laboratory [113]. This device provides electrical magnification of the output signal so that the pen motion is expanded by a factor of, say, 5, 10, or 20, as desired. An increase in signal of this magnitude is useful only if the noise level is kept low, however, so it is necessary to increase the slit width by an amount equal to the square root of the expansion factor and to decrease the gain to limit the noise level. Ordinate expansion is used most effectively over short intervals of wavelength, since it is extremely difficult to match cells sufficiently well to allow extended scans.

A method for accurate quantitative analysis now suggest itself. Suppose we are to analyze for approximately 40% A in B. We prepare standards of 38%, 39%, 40%, 41%, and 42% *A*. With the unknown in the sample beam, and each of the standards successively in the reference beam, we obtain a series of curves showing only the *difference* between the unknown and the standards. By using pen expansion, we can magnify the difference so that small differences in composition are reflected as large differences in the spectral trace (Fig. 6.6*a*). If, for example, we wish to use a 10× pen expansion factor and also decrease the normal noise level by a factor of 5, we need to increase the mechanical slit by a factor of $\sqrt{10} \cdot \sqrt{5} = \sim 7$ (and reduce the gain correspondingly). The ultimate accuracy, in fact, is reached when the slit width is increased to the point at which the monochromator exit slit image starts to spill over

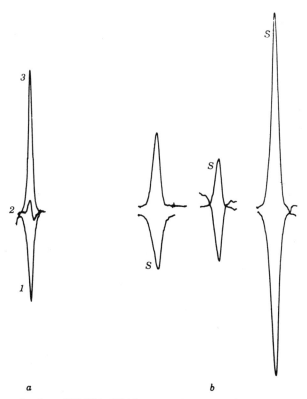

a   *b*

Fig. 6.6 Determination of [H(CH$_3$)SiO] in [(CH$_3$)$_2$SiO]$_x$ by difference spectroscopy at 5× pen expansion. (*a*) difference method. Trace 1, sample versus 2.82% standard; trace 2, sample versus 2.98% standard; trace 3, sample versus 3.19% standard. (*b*) Overcompensation method; *S* refers to unknown in the sample beam. Solutions are reversed for the other trace.

the edges of the detector [88, Chapter 7]. A calibration chart is conveniently prepared in terms of distance between the traces since the absorbance scale is meaningless when ordinate expansion is used.

Accuracy may be increased by another factor of two by using the overcompensation method [92]. The first scan is carried out as outlined in the preceding paragraph, but a second trace is made with interchanged samples. The cells are left in the same position. The difference then is twice that obtained for a single scan (Fig. 6.6b). Cell mismatch is automatically accounted for by this procedure. Difference spectroscopy is capable of giving accuracies on the order of 0.1% or better. As early as 1953, the difference method was used to determine cyclohexane in petroleum concentrates at the 85% level with a precision and accuracy of 0.1% [46].

As an alternative to preparing reference solution standards, Perry [83] suggests the use of neutral reference absorbers, such as screens. Calibrated rotating sectors [66] would also be suitable. If screens are used, it is important to:

1. Use screens having absorbance of 0.3 to 0.4 (40–50% transmittance). Higher absorbances are obtained by using screens in combination.
2. Treat the screens as optical elements; that is, protect them from dust and fingerprints.
3. Provide a mounting device such that the screens can be reproducibly inserted into the reference beam well away from a focal point.
4. Assure that the screens are not rotationally aligned with each other or with the slit.

The absorbance of each screen is measured carefully several times, using optimum conditions for quantitative analysis, and the average is taken as the correct value. Any residual error is compensated in the calibration curve. Screens are used in combination as well as singly.

Regardless of the reference absorber chosen, if ordinate expansion is used, standards for the analysis should be run each time the analysis is set up. Instrument settings must not be changed during the course of the determination.

A general procedure can be outlined for setting up analyses by difference spectroscopy.

1. Scan the spectrum of the sample and also of the component(s) sought.
2. Choose an analytical wavelength for minimum interference with the matrix.
3. Adjust sample thickness and concentration in solvent such that trans-

mittance of the analytical band is 30–40%. Reasons for this choice are discussed by Potts [88, Chapter 7].

4. Select instrument parameters, particularly slit width, that will give the required absorbance precision.
5. Prepare standards bracketing the expected sample concentration.
6. Scan over the analytical wavelength, with the unknown in the sample beam and successive standards in the reference beam.
7. If differences between traces are too small, determine how much ordinate expansion is necessary to achieve the desired precision.
8. Repeat step (5), using ordinate expansion. Increase the slit width by a square root of the expansion factor (and reduce the gain to give normal pen response).
9. Prepare a calibration curve and determine the unknown concentration.

Origin of Errors in Quantitative Analysis

It is the purpose of this section to point out some of the possible sources for errors in quantitative IR analysis and to suggest methods of minimizing them. In some cases, the magnitude of each contribution to the total error may be estimated, at least for the simple case of a single-component determination. Precision expected from multicomponent analyses is more difficult to evaluate. In the following discussion we assume that obvious mistakes are avoided and that uncertainties in weighing and glassware calibration are negligible.

The thickness of a cell can be determined quite accurately from its fringe pattern (see p. 117), but the determination is subject to some uncertainty if carried out in a casual manner or on an inaccurately calibrated spectrometer. Suppose, for example, that a 0.100-mm cell gives a pattern with 20 maxima at 835–1835 cm^{-1}. On a linear-frequency abscissa, a shift of the paper or a constant calibration error of, say, 5 cm^{-1} will not affect the calculated thickness. A nonlinear calibration error, however, of +5 cm^{-1} at 1835 cm^{-1} and −5 cm^{-1} at 835 cm^{-1} will result in a calculated thickness of 0.101 mm, or a 1% error. On the other hand, with a linear-wavelength presentation, misplacement of the paper by 0.025 μm will cause a 0.7% discrepancy in cell thickness, and a nonlinear calibration error of +0.025 at 5.45 μm and −0.025 at 11.98 μm will give a 1% thickness change.

Sometimes the apparent thickness of a cell varies from one spectrometer to another. This effect may be noted if the cell windows are not flat and parallel, because if the spectrometer radiation beams intersect different portions of the cell, the path length of the lens- or wedge-shaped cavity will in fact differ as one moves from the center toward the edge of the window. Well-made cells with windows flat to three or four wavelengths of sodium light do not show this effect.

It is highly desirable to work in regions of negligible solvent absorption. If solvent bands cannot be avoided, errors may be expected from inexact compensation of solvent bands. The effect becomes more serious for concentrated solutions and for regions of strong solvent absorption [42].

Even if the solvent does not absorb, the transmission of the sample and reference cells is likely to be different for reasons to be discussed. Correction is often made, particularly with the cell-in–cell-out method, by measuring the difference in absorbance of the two cells at each analytical wavelength when both are filled with solvent. The correction is subsequently applied to the sample absorbance, after accounting for the difference in transmission of the sample cell when it is filled with pure solvent as against sample solution at some nonabsorbing wavelength. It has been the author's experience, however, that even this procedure is not adequate to correct for cell mismatch when the sample cell windows become dirty, and the only means of getting reliable data is then to replace the sample cell so that the mismatch is small. The magnitude of cell difference error as a function of transmission has been evaluated by Martin [75].

Unless the refractive index of the solvent matches exactly that of the window material, some reflection loss will take place at the solution–window interface. A constant reflection loss is not in itself serious; a rapidly changing reflection loss is important, and this situation can occur under several circumstances.

It is well known that the refractive index of a liquid changes rapidly in the vicinity of a strong absorption band. If liquid samples are run undiluted or in concentrated solutions, changes in reflectivity are almost certain to be appreciable in some regions of the spectrum. Figure 6.7 illustrates the change in refractive index of chloroform in the vicinity of its 760-cm^{-1} band [50] [cf. equation (4.1)]. Refractive-index problems may be minimized by using solution concentrations of 2% or less [42]. The same consideration applies to solvents and provides an additional reason for avoiding solvents with background absorption, even though the bands are presumably compensated by pure solvent in the reference beam. Also, the sensitivity of reflection coefficients to trace deposits or fogging at the window–liquid interface is significant, and these artifacts are not always visible to the eye.

It is not widely appreciated that the refractive-index difference between solution and cell window noted in the preceding paragraph can introduce fringe patterns in the spectrum. Such patterns are rarely recognized but may lead to nonreproducible measurements. This effect is illustrated in

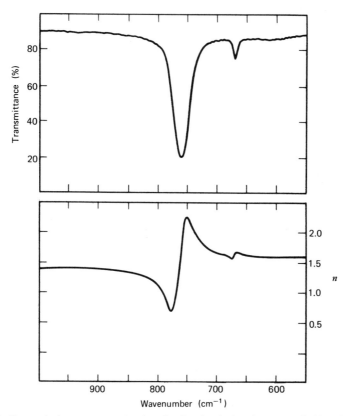

Fig. 6.7 Transmission spectrum (top) and refractive index (bottom) of chloroform, 550–1000 cm⁻¹. Reproduced, with permission, from *Spectrochimica Acta* [50].

Fig. 6.8, where the intensity of the absorption band plainly varies depending on whether its maximum corresponds to a maximum or a minimum in the fringe pattern. The baseline points will be affected in a similar manner, so that the cumulative effect may be quite appreciable. The effect has been evaluated by White and Ward [115], and some of their data are shown in Table 6.1. Obviously, the greater the difference in refractive indices, the greater the accentuation of fringe problem. Indices for some common window materials and solvents are listed in Table 6.2 from which the magnitude of the fringe problem may be estimated for several solvent–window combinations.

Effects of the fringes may be eliminated by (preferably) using a solvent that matches the refractive index of the cell windows, by intentionally

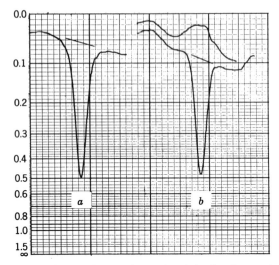

Fig. 6.8 Normal absorption band (*a*) and same band (*b*) showing result from background of interference fringes (exaggerated for clarity).

fogging the cells, or by using a spacing sufficiently large so that fringes do not appear.

Another more subtle effect of refractive-index mismatch is a small shift in absorption maximum (≤ 1 cm^{-1}). The shift occurs because the reflectivity of the cell window changes with the refractive index of the solution, and if solution and window have different indices, the reflectivity (and, therefore, the transmittance) will vary on each side of the absorption [123].

Convergence of the sample beam does not ordinarily affect the cell-

TABLE 6.1 **Fringe Amplitude Resulting from Cell–Solution Mismatch**[a]

n_{window}	n_{solution}	Reflectance	$\Delta I/I$
1.5	1.427 or 1.577	0.00062	0.025%
1.5	1.397 or 1.610	0.00125	0.50%
1.5	1.357 or 1.659	0.0025	1.0%

[a] Data from White and Ward [115].

TABLE 6.2 Refractive Indices of Window Materials[a] and Solvents[b]

Wavelength (in μm)	NaCl	KCl	KBr	CsBr	CCl_4	CS_2	C_2Cl_4	$HCCl_3$	C_6H_{12}
2	1.526	1.475	1.538	1.670	1.447	1.582	1.48	1.43	1.41
3	1.525	1.474	1.536	—	1.442	1.575	1.48	1.43	1.40
4	1.522	1.472	1.535	—	1.438	1.566	1.48	1.43	1.42
5	1.519	1.470	1.534	1.666	1.438	1.548	1.47	1.42	1.41
6	1.515	1.468	1.533	—	1.431	1.450	1.47	1.42	1.41
7	1.511	1.465	1.532	—	1.428	1.783	1.47	1.41	
8	1.507	1.463	1.530	—	1.422	1.657	1.46	.37	
9	1.501	1.460	1.528	—	1.406	1.636	1.45	.41	
10	1.495	1.456	1.526	1.662	1.384	1.628	1.42	.40	
11	1.488	1.452	1.524	—	1.339	1.622	—	.36	
12	1.480	1.447	1.522	—	1.169	1.618	1.50	.26	
13	1.471	1.442	1.519	—	—	1.615	—	—	
14	1.461	1.437	1.516	1.656	1.7	1.614	1.54		
15	1.451	1.431	1.513	1.655	1.60	1.612			
16	1.440	1.425	1.510	1.653	1.559	1.610			
18	—	1.414	1.500	1.648	1.533				
20	—	1.390	1.492	1.642	1.525				
25	—	—	1.463	1.625	1.497				
30	—	—	—	1.608	1.486				
35	—	—	—	1.596	1.494				

[a] Quoted by S. S. Ballard and J. S. Browder, in Smith [102].
[b] Data from Smith [102].

239

path length, but if a beam condenser or microscope attachment is used, the effect may become noticeable. Blout, Parrish, et al. [17] have discussed this case quantitatively and derived correction factors to be applied to the observed absorption that depend on the convergence angle, refractive index, and measured absorption.

A related effect that may result from thick cell windows or samples is a change in focus of the converging beam. The effect on the spectrometer is indeterminate, but the intensity, slit function, and stray light are presumably affected. The use of cell pairs with approximately equivalent window thickness is thus recommended.

Many modern IR spectrometers are thermostated at temperatures above ambient to prevent damage to water-soluble optical components from humid air. Absorption cells when mounted in position are also subject to this warming effect and, in addition, are exposed to a fairly intense beam of IR radiation. The result is that the temperature of the solution may change by several degrees, depending on how long the cell is kept in the spectrometer. The temperature dependence of the absorption bands is variable, but it is usually much greater than expected from density changes. For the CH_3 deformation in methyl esters, it amounts to about 0.2% in absorbance per degree centigrade [33]. Bands of associated molecules may show very large temperature dependence. The hydroxyl band in one study [54] was found to change absorbance by 0.01 unit per degree temperature rise, or by 40% during the temperature rise (from 24.5°C to 37°C) noted in the cell compartment of a commercial instrument. Cell thickness may also change with temperature. Quantitative analysis carried out with expectations of high accuracy obviously demands a well-thermostated cell.

The effects of certain well-known deficiencies of IR spectrometers on quantitative analyses have been extensively studied and the magnitude of the resulting errors evaluated. These factors include uncertainties in the transmittance, zero, and I_0 due to noise and stray light, tracking errors originating from the finite response time of the servo system, and resolution limitations.

The problem of noise has been somewhat alleviated in modern dispersive spectrometers by the use of more efficient optical systems, amplifiers, and detectors, but it still must be reckoned with when accurate transmittance values are to be obtained. As we noted earlier (pp. 40–45), noise may be reduced by either increasing the time constant of the servo loop or by widening the slits to increase the energy transmission of the spectrometer. The latter course is more practical but introduces another difficulty in that resolution is lost so that peak intensities are reduced. In quantitative analysis, however, reproducibility is more important than absolute absorbance accuracy, so that general rule is to widen the slits (and reduce the gain setting correspondingly) until an acceptable noise level is reached [85; 88, Chapter 6]. This level may be ±0.1% or less for best quantitative work.

Uncertainties in determining the zero, whether from noise or stray radiation, will influence the accuracy of the absorbance measurement. Similarly, errors in the estimation of I_0 are reflected in the absorbance readings. These factors have been considered quantitatively by Robinson [91], from whose paper Fig. 6.9 is reproduced, and by others [62, 75]. Fluctuations in the radiation source or the amplifying system also may contribute to uncertainties in the transmittance readings [75]; the net effect will be the same as that resulting from noise.

As a result of these and similar investigations, it has been generally accepted that, to minimize errors, one should work with transmittance readings in the range of 20–60%. It is obvious from Figs. 6.5 and 6.9 that the chances for error increase drastically below 10% and above 80% transmittance, and even rough semiquantitative analysis should not be undertaken near the limits of the transmittance scale. Sample emission is likely to contribute significantly to the error at high sample absorbance [70].

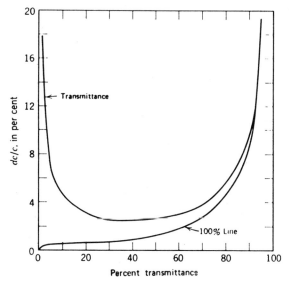

Fig. 6.9 Percent error in concentration for 1% transmittance error (*upper curve*) or 1% error in 100% line (lower curve). Reproduced, with permission, from *Analytical Chemistry* [91].

Double-beam spectrometers using the optical null system are subject to transmittance errors arising from nonlinearity of the optical attenuator [97], although good-quality spectrometers show errors of less than 0.4% transmittance [66]. This effect is not detectable by a Beer's law test and will not be important for analyses carried out on a single spectrometer, particularly if the same portion of the comb is used. Amplifier linearity is important in double-beam ratio-recording spectrometers.

The concept of "tracking error" or "dynamic accuracy" refers to the ability of the servo system or recorder pen to respond to instructions from the detector. Since all spectrometers have a finite time constant, the actual record only approximates the spectrometer signal [89]. Tracking error may be minimized by proper setting of the spectrometer controls and by the use of slow scanning speeds. It is wise, in fact, to scan the analytical bands at several speeds when setting up an analysis. The absorbance of a given band will show a dependence on scan time at the faster speeds, but a rate will be found that gives a maximum absorbance figure. The scan rate adopted for the analysis should be slower than this rate.

Other instrumental factors such as wavelength shifts arising from ambient temperature changes and atmospheric absorption are occasionally observed. The remainder of this section is devoted to less frequently recognized but nevertheless important factors influencing spectrometer accuracy.

It is generally known that gross misalignment of the spectrometer optics leads to inferior performance because of lack of optical purity of the signal and because of reduced signal:noise ratio. Further, however, more subtle changes such as warpage of the source or slight repositioning of the mirrors may produce no appreciable loss of resolution, but they may noticeably affect the long-term reproducibility of the spectrometer. This change occurs because the sample and/or reference beams then follow a slightly different path through the spectrometer, with a resulting effect on the slit function, stray light, I_0 line, and wavelength match of the sample and reference beams. Every optical adjustment should be followed by a check on the analytical curves used with that spectrometer, and IR sources, particularly Nernst glowers, should be checked periodically for warpage.

Any measurement made on a double-beam spectrometer in regions of solvent or atmospheric absorption will require longer scan times or wider slits than normal because of the energy loss from both beams. Records made in the water-vapor region often appear noisy because of air movement through the sample compartment, and no amount of slit widening will eliminate the noise. The only remedy is to enclose the compartment and allow it to equilibrate before recording the spectrum.

Other common sources of error include thermal radiation and reflection by the attenuator, chopper, slit jaws, and sample or reference materials. These factors contribute 0.1% or 0.2% uncertainty in the zero setting of a double-beam instrument [106], and 1% or 2% in a single-beam spectrometer.

The percentage error in absorbance resulting from cumulative 1% errors from each of the sources listed have been calculated by Martin [75] as is shown in Table 6.3. In practice, of course, some of the errors will cancel, and their magnitudes will certainly be different from 1%. Nevertheless, the tabulation is useful for demonstrating the results of uncertainties in the IR determination.

Most of the errors enumerated in this section can be avoided or minimized by making certain that the spectrometer is in good condition and its parameters correctly adjusted, by preparing working curves from synthetic mixtures, by checking the curves frequently with samples of known composition, and by running both standards and samples in a constant reproducible manner. In short, good analytical technique is essential for quantitative IR spectroscopy.

CALCULATIONS

Single Component

In the simplest case, one or more constituents of the sample has bands free from interference, and quantitative calculations may be done using equation (6.5). One first determines the value of the absorptivity a with material of known concentration (usually pure) and uses this value to calculate concentrations of the constituent in the samples. If, for example, the absorbance of a band for a pure species is 0.312 in a 0.101-mm cell and a 2.17% solution,

$$a = \frac{A}{cc'b} = \frac{0.312}{(2.17)\,(100)\,(0.101)} = 0.01424 \qquad (6.7)$$

so the concentration of this species in a sample showing an absorbance of 0.305 in a 0.115-mm cell and 2.05% solution is

$$c = \frac{A}{abc'} = \frac{0.305}{(0.01424)\,(2.05)\,(0.115)} = 91\% \qquad (6.8)$$

TABLE 6.3 Percent Error in Absorbance for Individual Errors of 1%[a]

Type of Error	Percent Transmittance Measured								
	10	20	30	40	50	60	70	80	90
Measurement of I_0 (1%) noise	0.43	0.62	0.83	1.09	1.44	1.96	2.80	4.48	9.49
Measurement of I (1% noise)	4.34	3.11	2.77	2.73	2.89	3.26	4.01	5.60	10.55
Stray light	4.14	2.56	1.98	1.67	1.47	1.33	1.22	1.13	1.07
Cell inequality	0.43	0.62	0.83	1.09	1.44	1.96	2.80	4.48	9.49
Maximum total error	9.34	6.91	6.41	6.58	7.24	8.51	10.83	15.69	30.60

[a] Data from Martin [75].

If analyses of this type are done repetitively, using the same standard, it is convenient to lump together the factor (bcc'/A) for the standard; thus $a' = (bcc'/A)$, where a' is the reciprocal of the absorptivity a. We then have $c = (Aa'/bc')$, and the calculation is easily carried out.

If Beer's law holds, the accuracy of results calculated in this way is limited only by errors of measurement. If Beer's law does not hold (a common occurrence), concentrations will be only approximate, but a calibration curve (or a correction curve) prepared from known standards will assure maximum accuracy (see Fig. 6.10). Composition of multicomponent mixtures may be calculated by exact solution of simultaneous equations if Beer's law is valid; if not, approximate or graphical methods may be used.

Multicomponent Analysis: Bouguer–Beer Law Holds

When several components are present, at least some of the analytical bands are likely to show interference from underlying absorptions of the

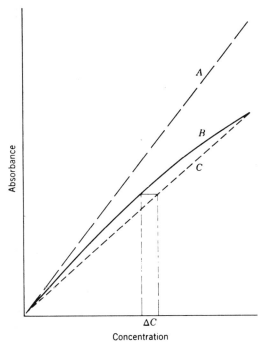

Fig. 6.10 Beer's law plots, showing curve for monochromatic radiation, A; observed curve using finite slit width, B; and curve assumed by use of single standard at 100% concentration, C. Error in this case is given by ΔC.

other constituents, although it is rare for all constituents to mutually interfere at all analytical wavelengths. It is, of course, advantageous to choose bands with a minimum of interference. In cases of overlapping bands, where associative effects are absent, the absorbance at a particular wavelength λ is given by the sum of absorbances of each absorbing species i at that wavelength, or

$$A_\lambda = a_1 bc_1 + a_2 bc_2 + \cdots + a_i bc_i \tag{6.9}$$

where the terms are defined as before, except $c_i = cc'$. Then for n components and N wavelengths, we have

$$A_A = a_{1A} bc_1 + a_{2A} bc_2 + \cdots + a_{nA} bc_n$$
$$A_B = a_{1B} bc_1 + a_{2B} bc_2 + \cdots + a_{nB} bc_n$$
$$\vdots \tag{6.10}$$
$$A_N = a_{1N} bc_1 + a_{2N} bc_2 + \cdots + a_{nN} bc_n$$

If one chooses strong bands with low or negligible interference for analysis, the diagonal terms $a_{1A} bc_1$, $a_{2B} bc_2$, $\cdots$, will be large and the other terms small or zero. The a values, of course, are determined from equation (6.5) by measuring the absorption of each component at all the wavelengths λ_N. The off-diagonal coefficients are subject to some uncertainty since these frequencies ordinarily do *not* fall at an absorption maximum. If one or more lie on the side of a sharp band, the coefficients are particularly vulnerable to frequency error. When the number of components is small, solution of the array of simultaneous equations (6.10) is easy. In the case $n = 2$,

$$c_1 = \frac{A_A a_{2B} - A_B a_{2A}}{(a_{2B} a_{1A} - a_{2A} a_{1B})b} \tag{6.11}$$

$$c_2 = \frac{A_B a_{1A} - A_A a_{1B}}{(a_{2B} a_{1A} - a_{2A} a_{1B})b} \tag{6.12}$$

Obviously the exact solution of i equations becomes much more difficult as $i = 3, 4, 5 \cdots$.

The array of coefficients of $c_1, c_2, \cdots, c_i$, in equations (6.10) forms a matrix that may be solved in terms of concentrations to give an inverted matrix:

$$c_1 = k_{11} A_A + k_{12} A_B + \cdots + k_{1n} A_N$$
$$c_2 = k_{21} A_A + k_{22} A_B + \cdots + k_{2n} A_N$$
$$\vdots \quad \vdots \quad \quad \vdots \quad \quad \quad \vdots \tag{6.13}$$
$$c_n = k_{n1} A_A + k_{n2} A_B + \cdots + k_{nn} A_N$$

For a two-component system, equations (6.11) and (6.12) form this matrix. Where i equals three or more, it is generally easier to solve for the coefficients k_{ii} numerically rather than implicitly. Such solutions may be carried out using standard methods of matrix algebra; specific directions and examples are given elsewhere [4, p. 86; 13, pp. 403–412]. For two- and three-component systems that are analyzed repetitively, nomographs that greatly simplify calculations may be easily constructed [101]. A shortcut method of evaluating the off-diagonal coefficients, using one or two mixtures of known composition, has been proposed by Perry [83, 86].

Equations (6.10) also can be solved by successive approximations; if the off-diagonal terms are small (little interference), convergence of the solutions is rapid. The method assumes in the first approximation that interferences are negligible (i.e., that the coefficients a_{1B}, a_{2A}, $a_{3B} \cdots a_{nN}$, $n \neq N$ are zero), and the concentrations c_1, c_2, $\cdots$, c_i are found. The initial values of c_i are then substituted into equations (6.10) and a new value of c_1 obtained. The process is continued and the calculations repeated until the desired degree of accuracy is reached [4, method E-168-67].

In practice, analytical bands can often be chosen such that interference from the other constituents is negligible or relatively constant for one or more components. Analysis is then greatly simplified since the concentration of this constituent can be calculated first and corrections found from graphs or tables to be applied to the other absorbances. Thus, by doing the calculation in the proper sequence, one can carry out the entire computation with only simple subtraction of correction terms and avoid the more tedious algebraic and approximation methods.

Multicomponent Analysis: Bouguer–Beer Law Not Valid

If the Bouguer–Beer law does not hold but the law of additivity is not violated, calculations may be carried out by one of several methods. If concentrations do not vary widely, the curved Beer's law plot may be approximated by a straight line over a limited concentration range. Linear equations may then be used between these limits, as discussed in the preceding section. Over broader concentration ranges, interaction terms may be introduced. Coefficients can then be evaluated by a least-squares procedure [10].

In many cases, particularly where a large range of concentrations is encountered, satisfactory results will be obtained by a graphical successive-approximation procedure [30]. For n components, n^2 curves of absorbance plotted against concentration at n wavelengths are prepared.

One can quickly find uncorrected concentrations for each component and, from these results, use the correction curves to find successively better values for concentration. If mutual interference is not too large, the concentration figures should converge quickly to within experimental error.

Algebraic and graphical methods also may be combined where appropriate [13, pp. 414–417, 87]. Concentrations are first calculated as if Beer's law held, and then correction factors (plotted graphically against concentration) are applied.

Analysis without Standards; Secondary Standards; Normalization Procedures

Analysis without pure reference standards is possible if: (1) the number of components, n, is known; (2) analytical bands with small or negligible interference are available; and (3) at least n mixtures, varying as widely as possible in composition, are available. Such conditions often obtain with a series of distillation cuts. If two mixtures x and y of components 1 and 2 are available, we may write, as before, for frequencies A and B:

$$A_{Ax} = a_1 b c_{1x}; \; A_{Ay} = a_1 b c_{1y} \tag{6.14}$$

$$A_{Bx} = a_2 b c_{2x}; \; A_{By} = a_2 b c_{2y}$$

But also

$$c_{1x} + c_{2x} = 1 \; (or \; 100\%) \tag{6.15}$$

$$c_{1y} + c_{2y} = 1$$

We now have six simple equations in six unknowns ($a_1, a_2, c_{1x}, c_{2x}, c_{1y}, c_{2y}$), and solution is relatively easy.

As the number of components increases, the set of equations (6.14) and (6.15) also becomes more complex. Since the compositions necessarily become more similar, the accuracy of the determination decreases.

If mutual interferences occur, it is still possible to set up a system of equations to solve for the composition of binary mixtures, but in this case data from additional mixtures will be required because of the two additional unknown absorption coefficients [88, Chapter 6].

If pure standards are not available for determining absorption coefficients, the procedure described in the preceding paragraphs can be used to calculate values for a. Mixtures or even samples that have had an independent determination of the concentration of each constituent by some other method can be used. The accuracy of the IR analysis is then, however, totally dependent on the independent analysis. Furthermore, the risk of unknown or unidentified (and possibly interfering) components being present makes the use of secondary standards unattractive.

Results for multicomponent mixtures are sometimes normalized to 100%, but this should not be done routinely. The practice is justified only if all components are affected equally by some variation of unknown magnitude such as dilution error or uncertainty in cell thickness and if it is established that no additional species are present in the mixture. The sum of the concentrations of all components provides a good check on both the precision of the analysis and the unexpected appearance of unrecognized constituents.

Statistical Evaluation of Results

Every routine analytical procedure should be subjected to statistical analysis. This procedure not only defines the probable limits of error in the analysis, but also helps pinpoint the source of these errors. Details of the procedure are given elsewhere [3], so only a brief elementary discussion is given here.

We first distinguish between the *precision* or reproducibility of a measurement and its absolute *accuracy*. Estimates of standard deviation made using difference values from duplicates analyses refer to precision; estimates made from the difference between assumed and found values of synthetic standards give a measure of accuracy. Both types of data are useful. It is well known that 67% of the values of a replicate analysis fall within ± 1 standard deviation or $\pm \sigma$, 95% within $\pm 2\sigma$, and 99+% within $\pm 3\sigma$. The true standard deviation σ is distinguished from the estimate of standard deviation, s. For replicate measurements, if x represents the observed value of the ith measurement.

$$s = \sqrt{\frac{\Sigma x_i^2 - (\Sigma x_i)^2/n}{n-1}} \qquad (6.16)$$

or if all sets are duplicates and d is the difference between pairs,

$$s = \sqrt{\frac{\Sigma d^2}{2n}} \qquad (6.17)$$

If s varies approximately in proportion to the value measured, *the coefficient of variation* or *relative standard deviation* = $s/$(true value) provides a convenient comparison parameter. To illustrate, a portion of a statistical study involving measurement of an absorption band in a solid by two operators over a 2-month period is shown in Table 6.4 (a). From the results one can obtain an estimate of the probable error of measure-

TABLE 6.4   Replicate Concentration Measurements

	Operator No. 1	Operator No. 2
a. Measurements on Single Solid Sample		
	0.640	0.651
	0.652	0.640
	0.635	0.636
	0.643	0.632
	0.638	0.627
	0.638	0.629
	0.645	0.633
	0.648	0.638
	0.627	0.630
	0.643	0.633
Mean =	0.641	0.635
s =	0.0070	0.0069
Coefficient of variation =	1.09%	1.09%
b. Measurements on Sample Run as Solutions		
	85.3	81.2
	84.3	82.2
	81.5	81.5
	80.3	80.5
	80.3	78.2
	85.0	82.2
	83.5	84.5
	84.1	84.8
	80.5	85.8
	84.6	83.4
Mean =	82.94	82.43
s =	2.06	2.64
Coefficient of variation =	2.48%	3.20%

ment, including instrumental errors, of each operator and compare the errors (by another test) to see whether the difference is significant. By repeating the process with solution samples, one can add the weighing and dilution steps as in Table 6.4 (b). Other analytical operations may be treated similarly. One may thus arrive at some conclusions as to the uncertainties in each step of the analysis.

Probably the most meaningful data are obtained through the use of "blind" samples, that is, the same material submitted under code numbers. The use of coded samples helps preclude conscious or unconscious bias in the data gathering.

SPECIAL PROBLEMS AND TECHNIQUES

Quantitative Gas Spectra

In most cases quantitative analysis of gas mixtures will be more easily carried out by GLC or by mass spectrometry, but IR can also be used successfully, provided that due attention is given to proper standard preparation and to the phenomenon of pressure broadening. Effects of the latter may be minimized by pressurizing all gas samples to 760 mm with an inert gas such as nitrogen. Sorption of vapors of some compounds within the cell can be minimized by preconditioning the cell with sample.

Quantitative infrared vapor spectroscopy has been discussed by Friedel and Queiser [41], who also describe a convenient cell-filling technique for use with volatile liquids. Materials that are gaseous at room temperature can be handled in the usual vacuum manifold system. Very dilute gas mixtures can be prepared by a diffusion technique [78].

Solids Analysis

It is the opinion of a number of workers in the field, including the author, that quantitative analysis on solid materials as such should be avoided. Nevertheless, at times no choice is possible and, if due care is taken, satisfactory results can often be obtained by a quantitative mull or KBr pellet technique.

The difficulty in analysis of powders is the dependence of absorbance on sample uniformity. It has been pointed out by Jones [65] that if a sample with a band showing a true absorbance of 1 has only 10% of clear space in the sample beam, the apparent absorbance will be 0.775. Errors for other ratios of transmitting to absorbing areas are given, and it is emphasized that the effect becomes rapidly greater as the absorbance increases. This effect is called the "mosaic" effect, and its magnitude depends on particle size, shape, and distribution. As the concentration of particles becomes large, the clear space (and the magnitude of the error) diminishes [63]. Another factor, often unrecognized, is that band intensity for crystalline materials depends on particle size. A study of the crystalline solid chloranil showed that as the particle size was varied from 12 μm to 160 μm, the absorptivity of several bands (in the KBr matrix) decreased by a factor of as much as 4 (Fig. 6.11). A similar effect has been noted with quartz [111]. Bands may show an apparent shift in frequency as well as intensity. The reason is that surface rather than bulk modes are being observed, and surface modes are sensitive to the dielectric constant of the surrounding medium [94]. It is, therefore, apparent

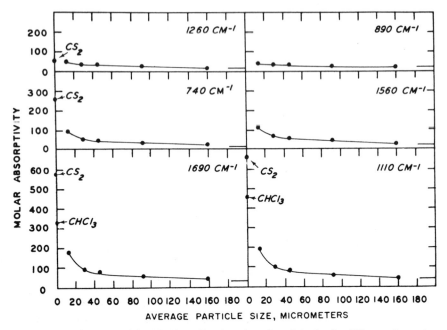

Fig. 6.11 Molar absorptivity of chloranil as function of particle size for different absorption bands. Reproduced, with permission, from *Applied Spectroscopy* [98].

that incomplete distribution of the absorbing species in the beam, sample particles that are too large, or variation of particle-size distribution from one sample to another, will result in anomalous band intensities. It is usually recommended that the particle diameter be less than the shortest wavelength of light used (2 μm in most cases). Thus, if solution spectra are impractical, in order to do meaningful quantitative work with KBr pellets or mulls, one should make certain that the sample is consistently pulverized to the required degree of fineness.

A number of quantitative methods have been published, however, in which a KBr matrix is used. The success of such a technique obviously depends on reproducibility in grinding and dispersing the samples.

Quantitative mull spectra can be obtained by *internal standardization*. This term refers to the practice of mixing a known amount of noninterfering material having an easily measured absorption band with the weighed unknown, and comparing the ratio of band absorbances. Such a ratio is independent of sample thickness. Some of the materials proposed for use as internal standards are calcium carbonate, sodium azide, d-alanine, lead thiocyanate, naphthalene, and potassium thiocyanate. Of these, the inorganic materials have the simplest spectra, and the thiocyanates show a strong absorption at 4.8 μm that is usually free of interference from the sample.

Applications of internal standardization include analysis of poly(vinyl acetate-chloride) polymers for acetate content [116]; determination of the free-acid content of aluminum soaps, run as KBr pellets [116]; and analysis of isomeric phthalic acids run as mineral-oil mulls [19].

Quantitative Analysis of Polymeric Materials

The quantitative analysis of polymers presents some peculiar problems, and in many cases the approach must be tailored to the problem at hand. Most of the difficulties arise from the insolubility of polymeric materials in common IR-transmitting solvents and from the uncertainties inherent in preparing standards. Some ingenious solutions to these problems have been devised and are described briefly.

Satisfactory standards may be obtained in some cases by conventional copolymerization methods, but often the resulting compositions are not known precisely enough for use as standards. Chemical methods of analysis, if available, are useful for characterizing polymer composition. A third method of standardization is by the use of radiotracers. In one example, C^{14}-labeled ethylene was used to prepare ethylene–propylene copolymers, and the compositions were determined from the specific activities [34]. In another case, standards for IR analysis of a methyl isopropenyl ketone–butadiene–acrylonitrile terpolymer were prepared using methyl isopropenyl ketone–C^{14}. The acrylonitrile content was determined by using the Dumas method for nitrogen [105]. Nuclear magnetic resonance has also been used to establish the composition of standards. Tosi and Ciampelli [109] have reviewed methods of standardizing ethylene–propylene copolymer compositions, and Hampton [47] has tabulated literature references to 39 literature methods relating to quantitative analysis of polymers.

Infrared examination of the polymer may be carried out in several ways. Most common is the technique of casting a thin film of the material from solvent. Although the thickness is not uniform nor accurately known, it can be eliminated from the calculations by using the absorbance ratio of two bands in the spectrum, one from each component. The method can be applied to a copolymer with any number of components, provided that a unique absorption band can be found for each of them. Peak transmittance values should be optimized as shown in Fig. 6.5.

Polymers containing fillers and plasticizers are often sampled by solvent extraction [47]. Plasticizers may be soluble in mild solvents such as CS_2 or ethyl ether and can be extracted from the fragmented polymer by using a Soxhlet apparatus. A CS_2 extract can be run directly in the IR spectrometer. The polymer is separated from the filler by use of a stronger solvent such as o-dichlorobenzene. The polymer may then be cast as a film and the filler examined by a mull or KBr pellet technique. An example of this procedure as applied to the quantitative analysis of polyvinyl chloride compounds has been given [21].

Sampling by ATR is discussed later (p. 256).

Another technique, applicable to insoluble as well as heavily filled polymers, is controlled pyrolysis [15]. In the example given, buna-n-phenolic copolymers were heated at 540–580°C in vacuum and the pyrolysate collected in a dry-ice trap. Analysis of the trapped material gave results for the phenolic component reproducible to within ±2% in the 0–33% range, and the amount of noncombustible filler could also be estimated. Analysis of specific polymers is discussed more fully in other works [47, 110], which include citations to papers on many different types of polymers as well as to quantitative infrared methods for polymers.

An interesting approach to semiautomatic analysis of polymer films has been given by Johnson, Cassels, et al. [64], who have linked a commercial spectrometer with a computer for unattended analysis of as many as 72 consecutive samples.

Quantitative Analysis of Associated Species

As we have seen, intermolecular association is strongly concentration dependent. Furthermore, the vibrational frequencies and absorption intensities of the molecules involved in association usually change with concentration. Also, individual species of a mixture may associate with themselves, with the solvent, or with other components in varying proportions, depending on their relative concentrations. This type of association occurs when the molecules involved have acidic and basic sites (in the Lewis sense).

Primary aromatic amines furnish a good example of these associative effects. In a study of the first overtone of the NH stretching vibration, it was found [114] that the peak molar absorptivity of the NH symmetric stretch varied from 1.87 in cyclohexane to 0.61 in acetonitrile, with other solvents giving intermediate values. This effect, incidentally, is the reverse of that reported for the corresponding fundamental band. A change of solvent in such a system would clearly demand complete restandard-ization of the working curve.

The hydroxyl band furnishes another well-known example of concentration dependence. It frequently is desirable to measure the concentration of either the hydroxyl group or a molecule containing this group by spectroscopic methods. Although this can be done sometimes by using one of the techniques discussed in the following three paragraphs, because of the tendency of hydroxyls to hydrogen bond to oxygen and other electronegative atoms (see pp. 155–161), IR spectroscopy is not as useful in this case as it is for other quantitative analyses. If the solvent and other molecular species present are essentially inert, the relative intensities of the associated and unassociated OH bands will be determined by concentration and temperature of the solution, with more dilute solutions favoring dissociation. Contaminants (e.g., water) or other protondonating or -accepting species will change the dissociation equilibrium and the relative intensities of the two absorptions. Certainly, then, neither band will follow the Bouguer–Beer law, and attempts at quantitative analyses are frustrated. Furthermore, as has been implied, any other group involved in association may show similar effects to a greater or less extent, so one must exercise some care in choosing frequencies for quantitative analysis when possibilities for association exist. A sound knowledge of group frequencies and their origin will prove invaluable in making this choice.

In some cases special methods can be used to circumvent these difficulties. Perhaps the most common of these is to dilute the sample, using an inert solvent, to at least 0.01 molar, where the concentration of associated species is negligible. A longer than normal absorbing path is then necessary, so a highly transparent solvent must be used. In the short-wavelength end of the IR region, CCl_4 fulfills this criterion.

A second method of dealing with association effects is to ignore them. This tactic is sometimes successful, but only if the system itself is relatively constant in composition and interfering contaminants are either absent or can be corrected for. Thus the hydroxyl number of alkyd resins has been determined from the absorbance of the associated OH band of the resin in CH_2Cl_2 solution. The result is corrected for the acid number of the resin and for its water content [80]. Obviously, such an empirical method requires a different calibration curve for each type of resin, and frequent checks on the accuracy of the determination must be made.

The third approach to analysis of associated species is to run them under conditions of complete association. This can be done by adding a Lewis base (or acid, as appropriate) as solvent or, possibly, by using the material as its own solvent. Polypropylene glycols, for example, can be analyzed *in situ* for hydroxyl [22], since the —OH group forms an intramolecular hydrogen bond with the ether oxygen and the resulting absorption band follows the Bouguer–Beer law quite accurately. An independent water determination may be necessary to correct the absorbance value obtained for the —OH group. In the near IR, chloroform is often used as an associative solvent to tie up hydroxyl groups and ensure more reproducible analysis. Polyesters and polyethers have been analyzed for hydroxyl number in the 2–3.2-μm region, using CCl_4 containing 10% $HCCl_3$ as a solvent [54]. Mixtures of H_2O and D_2O can be analyzed using the ratio of the 2560 cm^{-1}:3530 cm^{-1} absorptions [36]. Absorption bands of associated species are particularly sensitive to temperature fluctuations, and if such bands are being used for analysis, the sample temperature should be controlled carefully.

In summary, then, it is best to avoid quantitative measurements on frequencies of groups involved in association. If such measurements are necessary, one should keep conditions as invariant as possible and make frequent calibration checks.

Group Analysis

The direct determination of chemical genera, such as aldehydes, acids, and alcohols, without regard for the individual molecular species present, furnishes an interesting example of the application of group frequencies to quantitative analysis. Such determination of group concentrations is of interest in the petroleum and chemical industries. In one study [95, 96], alcohols, acids, aldehydes, ketones, esters, and ethers were determined by measurement of their absorptions at 3635 cm^{-1} (2.75 μm), 3550 cm^{-1} (2.82 μm), 2720 cm^{-1} (3.68 μm), 1720 cm^{-1} (5.8 μm), 1140 to 1300 cm^{-1} (7.7 to 8.8 μm), and 1060–1220 cm^{-1} (8.2–9.4 μm), respectively. In another study [61], *trans* unsaturation in fats, oils, and esters was monitored by IR.

The accuracy of these analyses depends on the constancy of the absorption coefficient for each group from one structure to another, and to a lesser extent on mutual interactions between members of different groups. In general, inaccuracies in group analysis are larger than errors in multicomponent analysis for individual molecular species and may be on the order of 10–20%. On the other hand, related groups such as methylene units in the *n*-paraffin hydrocarbons show quite uniform absorption coefficients independent of chain length [124], which presumably can be characterized with considerable accuracy.

Attenuated Total Reflectance

Quantitative analysis using ATR is convenient if the sample contains an internal standard, that is, an absorption for which the intensity does not vary as the composition of the sample changes. External standards can be used, but a number of complicating factors make this approach difficult [118]. First, the sample placement and the efficiency of contact between the reflection element and the sample must be highly reproducible. Second, other parameters—(viz., angle of incidence and optical alignment) must also be reproduced. The conditions can be met more easily for liquid than for solid samples, but here another pitfall may await, specifically, the tendency for segregation to occur in some samples, possibly because of electric field gradients at the surface of the reflection element. Thus the sample "seen" by the radiation will not be representative of the bulk. Sometimes one component of the solution is adsorbed on the surface of the element, making quantitative work out of the question.

Use of band ratios offers a more accurate method for quantitative analysis of solids. In a binary system, one band from each component is selected. The bands should be free of interference and not too far separated in wavelength for best results. A series of standards is run and a working curve prepared, which is used for analysis of unknowns. Reproducibility of sample placement and contact area is not so critical since both bands should be affected equally (although it would be well to verify this point). In one example, cotton–nylon blends in the 20–80% range were analyzed, using a 1-mm-thick KRS-5 plate with 60° angle of incidence [118]. The vinyl acetate content of ethylene–vinyl acetate copolymers has been monitored by using the ratio of the 1735 cm^{-1} carbonyl absorbance to that of the 1465-cm^{-1} CH band [38].

Regardless of how the analysis is carried out, the spectrometer must be adjusted to have adequate servo energy, or inferior results will be obtained.

Analysis for Trace Impurities

Although the IR method is not usually considered to be sensitive to trace impurities, a number of techniques have been used to extend its sensitivity into the parts-per-million range. In this section discussion is restricted to the determination of known minor components in large samples, as distinguished from analysis of microsamples, which is really a problem in technique. Several approaches have been used successfully for trace analysis by IR.

Long Sample Paths

Trial spectra of the matrix material at sample paths 10–1000 times longer than usual sometimes reveal "windows" in the absorption pattern in which bands of the impurity may be found. The C—H stretching region is often valuable in such problems. This technique, where applicable, is rapid, sensitive, and accurate; it should be the first approach investigated for any problem in trace analysis. It has been applied to the determination of 0–100 ppm moisture in the refrigerant dichlorofluoromethane [32]. Path lengths of up to 50 m of air using multipass absorption cells give sensitivity in the fractional parts-per-million range for many common atmospheric pollutants (Table 6.5) [59].

Difference spectroscopy combined with ordinate expansion offers an alternative approach to the use of thick samples if a suitable reference material is available. The sensitivity of dispersive spectrometers is then limited only by the width of the detector element, since the pen expansion ratio and sample thickness may be increased indefinitely as long as the slit is widened correspondingly to provide an acceptable signal:noise ratio.

Some restrictions on this technique that are not immediately obvious should be pointed out. First, the spectrometer must be in good alignment

TABLE 6.5 Sensitivity for Common Atmospheric
Pollutants in a 40-m Gas Cell[a]

Compound	Minimum Detectable Concentration[b]		
	cm^{-1}	(mg/m^3)	(ppm v/v)
$CH_2{=}CHCHO$	1720	0.05	0.02
H_2S	1290	70.	50.
HCl	2920	12.	7.
NO_2	1630	0.03	0.014
CO	2160	1.	0.8
C_2H_4	946	0.12	0.10
SO_2	1370	0.09	0.03
CH_4	1300	0.06	0.085
O_3	1050	0.11	0.05

[a] Data from Hollingdale-Smith [59].
[b] Minimum amounts detectable in the atmosphere may be somewhat higher because of interfering absorptions from other constituents.

and the cells well matched for transmission, or else a rapidly sloping background will result. Second, difficulty is likely to be encountered in regions of atmospheric absorption because of random air currents. Enclosing the sample compartment and flushing it and the spectrometer with dry gas will minimize this problem. Third, it should not be assumed that if, say, 10 ppm of a material is detectable at $5\times$ pen expansion, 2.5 ppm can be seen at $20\times$. Such a proportional relationship very often does not hold because of loss of resolution from slit widening, increased background slope, and the increased difficulty of exactly compensating the major constituent.

Difference spectroscopy has been used to determine minor components in detergent products [16], six components of a C_4 gas mixture [84], and in the study of branching in polyethylene [119, 120].

Interferometer spectrometers utilize multiple scans and special smoothing routines to improve the signal:noise ratio for ordinate expansion. Compensation of the lattice bands of a 2-cm-thick sample of semiconductor silicon with a purified reference specimen permitted detection of as little as 7 ppb of oxygen in the sample slug [67].

Infrared techniques have been applied to the solution of many difficult pollution and environmental problems. On-site analysis, of stack gases, for example, can be performed at extremely low levels with interferometer spectrometers [57]. The highly toxic substance nickel carbonyl has been determined at less than 1 ppb in the presence of a 10^6-fold excess of CO, which interferes [74]. An interesting approach to the analysis of hexafluoroacetone (HFA) in air has been taken by Baker and Carlson [7]. They attain high analytical sensitivity by reacting the HFA with K_2CO_3 and H_2O to give HCF_3, which can be detected at much lower concentrations than HFA. Detection limits for many contaminants can be lowered to the 0.05–1-ppm range by drying (with a special drier) and compressing the air sample. The spectrum is scanned at 20 m path length and 10 atm of pressure [6]. Solar spectra in the region 800–1250 cm^{-1} have been analyzed using computer simulation to determine 110 ppt (parts per trillion) CCl_2F_2 (ground level equivalent) in the atmosphere [81].

Infrared lasers have found application for analysis at low levels of nitrogen dioxide [40], vinyl chloride [39], and various other gases [45]. The best frequencies for analysis of O_3, N_2O, CO, CH_4, H_2O, and CO_2 using narrow-band IR sources such as lasers have been determined [44]. Parts-per-billion sensitivities are achieved with tunable lasers using cooperative reflectors [55, 56].

Spectroscopic methods for air-pollution measurements have been reviewed by Hanst [48, 49].

Correlation Spectroscopy

Gas analysis in the parts-per-billion range can be enhanced by a sensitive analytical technique called *correlation spectroscopy* [117]. In this technique the spectrum of incoming radiation is correlated point-by-point with the spectrum of a target gas that is stored in the spectrometer. In essence, the sample spectrum is multiplied by the target spectrum and the result integrated. Regions where interference exist are selectively blacked out. Spectra can be observed in either emission or absorption.

Three systems for correlation spectroscopy are described. The first is a grating monochromator in which the exit slit is replaced by a mask that corresponds to the absorption pattern of the target gas. A second mask covers the same spectral region but is designed to transmit only background radiation adjacent to the absorption bands. The radiation is alternated between the two masks, and the magnitude of the detector signal is a measure of the intensity of the target gas spectrum in the incoming radiation.

The second system is similar to the nondispersive gas-filter analyzer (see pp. 270–272), in which the target gas is used as the sensitizing gas. Such a device has been described for use as an ambient-air CO monitor [26].

The third type of correlation spectrometer utilizes an interferometer. The target gas spectrum is stored in the instrument as an interferogram. Data processing is carried out easily with the spectrometer computer. Applications to NO, CO, SO_2, HCl, and HF in stack gases (by absorption) have been described [52]. Table 6.6 shows detection limits for common gases in the emission mode [58].

Concentration Techniques

Concentration of the sought-for impurity by physical or chemical means is a powerful method for trace determinations. A great deal of leverage on the original sample may often be obtained at the cost of an extra step in the analysis and the attendant risk of incomplete recovery of the impurity.

So many possible methods are available for separating minor constituents that only a few can be mentioned. Chromatographic techniques, solvent extraction, crystallization, and distillation have all been used for isolation of trace impurities. The combination of GC and IR spectroscopy has been discussed earlier. Paper chromatography has been used for separation of biological materials; a few micrograms suffices for an IR spectrum [108]. Solvent extraction is a simple, effective technique of

**TABLE 6.6 Remote
Detection of Atmospheric
Pollutants in Emission**[a]

Compound	Sensitivity (in ppb)
CO	5
NH$_3$	10
CH$_4$	0.6
NO$_2$	2

[a] Path length 5 m and source temperature 200°C; data from Hirschfeld [58].

wide applicability. Extraction of waste water with CCl$_4$, for example, permits determination of oils in the water to 0.1 ppm and phenols to less than 10 ppb [100]. Silicone fluids used as defoamers in food products have been determined in the low parts-per-million range by extraction with CS$_2$ or other solvent [60]. Fractional crystallization combined with difference spectroscopy was used to determine catechol and similar impurities in hydroquinone [9]. Solvent separations have been applied to the routine control analysis of polyethylene additives by first pulverizing the polymer and then extracting it with CCl$_4$ and CS$_2$ [104].

Indirect Methods

Minor constituents in a material may be reacted chemically to form derivatives or unique compounds that are susceptible to IR analysis. Water in inorganic liquids has been determined by reacting it with calcium carbide to form acetylene, which is measured with an IR spectrometer [37]. Sensitivites below 1 ppm are attainable, but the procedure is somewhat involved and special equipment is necessary. Water in somewhat greater amounts can be determined by treating it with 2,3-dimethoxypropane in acidic medium and measuring the carbonyl band of the acetone formed [29]. A novel example of the indirect method is the determination of traces of oxygen in sodium metal. The sodium is converted to NaCl in a Wurtz reaction with n-amyl chloride, and the sodium oxide is reacted with CO$_2$ to form Na$_2$CO$_3$. The NaCl matrix is pressed into a disk and the amount of carbonate is determined. A range of oxygen concentrations of 20–300 ppm is covered in this method [31].

Combination Techniques

Combining techniques such as IR with UV or mass spectroscopy, or with GLC or LC to give a combined analysis, is often advantageous. Many otherwise formidable problems, such as the identification of *o*-xylene oxidation products [20], have been successfully solved in this manner. The use of such methods for routine analysis is less common, but where one or more IR-insensitive components are vulnerable to determination by titration, polarography, or UV spectrometry, such an approach should not be overlooked. The important consideration is the relative ease, speed, and accuracy of the combined analysis versus the IR method.

Examples

The following examples of quantitative analysis have been chosen from the literature to illustrate the applicability of IR analysis to a broad range of problems. Conditions used in these analyses are not necessarily optimum—better results could have been obtained in some instances by use of the practices discussed earlier in this chapter. Many of the best examples of quantitative IR analysis have never been published, as they involve proprietary materials or methods.

   a. Measurement of 0.1 to 0.8% 4-methyl-2,6-di-*tert*-butylphenol in mineral oil (5): The phenol was used as an oxidation inhibitor, and a rapid, accurate quality control method was needed. Since the hydroxyl group is hindered by the bulky *tert*-butyl groups, it does not hydrogen bond at lower concentrations and can be used for analysis. An absorbance-ratio method has also been used for this analysis [27].

   Matrix: mineral oil.

   Sample form: liquid in 1-mm cell.

   Band measurement: OH at 3650 cm^{-1} (2.74 μm). Spectrum scanned 2.65–2.85 μm. Zero absorbance point measured at 2.82 μm (Fig. 6.12).

   Calculations: graphical; $A = A_{2.74} - A_{2.82}$ plotted against concentration.

   Precision: $s = 0.006\%$ in 0.2–0.8% range.

   Accuracy: $s = 0.01\%$.

   Time required: 5 min.

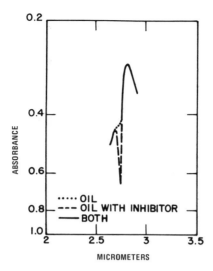

Fig. 6.12 Spectra of oil with and without inhibitor. Reproduced, with permission, from *Applied Spectroscopy* [5].

b. Analysis of five C_{10} aromatics [82]: Although the constituents of this mixture are volatile and hence amenable to analysis by GLC, the method is equally applicable to nonvolatile materials. Samples were run without dilution, and measurement of cell thickness was eliminated by normalizing the results (normalization could not be used if additional unmeasured components were present). Time required was about 45 min.

Matrix, analytical wavelengths in μm, (reference wavelengths), and relative values of absorptivity at each analytical wavelength:

Component	8.56 (8.79)	10.00 (10.53)	10.65 (10.53)	11.54 (11.40)	8.92 (8.79)
iso-Butylbenzene	0.870	−0.068	−0.074	−0.079	0.244
sec-Butylbenzene	0.031	0.529	−0.142	0.004	0.023
o-Diethylbenzene	0.002	−0.083	0.395	0.056	0.071
m-Diethylbenzene	0.215	0.053	−0.021	0.591	0.020
p-Diethylbenzene	−0.063	0.003	−0.013	0.064	0.760

Sample form: undiluted liquid in 0.1-mm cell.

Standards: pure (99+%) hydrocarbons.

Band measurement: absorbance at reference wavelength subtracted from absorbance at analytical wavelength.

Calculations: algebraic.

Precision: s not calculated; average absolute error 0.5%; average relative error 2.7%.

c. Determination of alcohol–water binary mixtures [72]: Analysis of 0–100% methanol–water, ethanol–water, *n*-propanol–water, and *iso*-propanol–water mixtures are described in this paper, but we examine only the latter. To avoid the use of thin transmission cells, multireflection ATR with a 45° germanium element was used. Somewhat surprisingly, the Bouguer–Beer law was obeyed even for the hydrogen-bonded groups.

Matrix: alcohol–water.

Sample form: liquid in ATR cell.

Band measurement: iso-propanol, 950 cm^{-1}; water, 3350 cm^{-1} (corrected for alcohol OH absorption).

Calculations: Graphical (Fig. 6.13).

Precision: $s \leq 4\%$ for average of seven determinations.

d. Analysis of polyester–wool mixtures containing 25–75% polyester

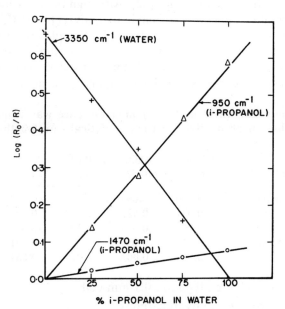

Fig. 6.13 Absorbance–concentration plots for *iso*-propanol–water mixtures, as determined by ATR. Reproduced, with permission, from *Spectrochimica Acta* [72].

[12]: A rapid and precise method was needed to characterize fabrics made from blends of polyester–wool fibers. Sampling was done by ATR, and since this is a binary system, an absorbance-ratio method could be used. Precision would have been better if wider slits and lower gain (to give a less noisy trace) had been used.

Matrix: polyester–wool fibers.

Sample form: cloth positioned on both sides of a 45° KRS-5 ATR element.

Band selection: wool, 1520 cm^{-1} (interference from the polyester is compensated for by using band ratios); polyester, 1714 cm^{-1} (Fig. 6.14).

Standard preparation: unprocessed wool and polyester cloth ground separately in a Wiley mill; humidity equilibrated; weighed; mixed in aqueous slurry; and dried.

Calculations: Graphical (Fig. 6.15).

Precision: s not calculated, but nine replicates gave results within 10% of the average composition.

e. Determination of HF, HCl, SiF$_4$, and HCF$_3$ in WF$_6$ gas [25]: One important method for production of high-purity tungsten metal is the hydrogen reduction of WF$_6$ gas. Impurities in this gas have a large effect on the properties of the metal, however, and must be closely controlled.

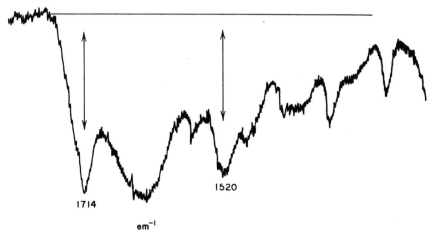

Fig. 6.14 Baseline and peak measurement of polyester and wool bands. Reproduced, with permission, from *Applied Spectroscopy* [12].

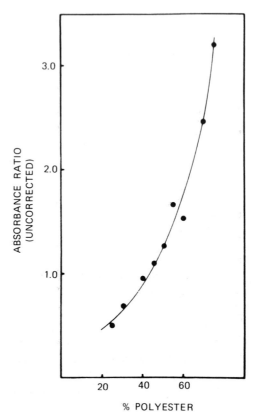

Fig. 6.15 Calibration plot: A (polyester): A (wool) versus percent polyester. Reproduced, with permission, from *Applied Spectroscopy* [12].

Analysis is somewhat difficult because WF_6 is extremely reactive; for instance, it reacts with moisture to form HF. A special sampling system which allowed flushing and dry N_2 purging of the infrared cell was used.

Matrix: WF_6 gas at atmospheric pressure.

Cell: 10-cm length, made from type 304 stainless steel with BaF_2 windows.

Band selection (see Fig. 6.16):

Compound	Frequency	Sensitivity (w ppm)
HF	3850 cm^{-1} (2.60 μm)	100
HCl	2860 cm^{-1} (3.50 μm)	100
SiF$_4$	1030 cm^{-1} (9.73 μm)	10
HCF$_3$	1150 cm^{-1} (8.71 μm)	10

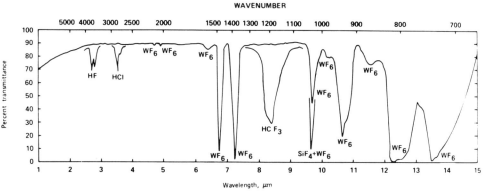

Fig. 6.16 Absorption spectra of WF_6 at ~1.3 atm and 25°C in 10-cm cell. Impurity absorption bands are also shown. Reproduced, with permission, from *Applied Spectroscopy* [25].

Standard preparation: standard addition of impurity gases into the cell using a gas syringe (Teflon syringe used for SiF_4).

Calculations: algebraic (nonlinear for HF and HCl).

Precision: s not calculated, but "best precision obtained in a routine analysis sequence was ±6% using duplicate samples." Most of this error was due to sampling, however, as repeat scans on the same sample were precise to about ±1%.

f. Quantitative analysis of surface $Me_3SiO_{1/2}$ and SiOH groups on treated SiO_2 [2]: The surface of silica gel and aerosil SiO_2 is composed of silanol (SiOH) groups and adsorbed water. The surface can be modified by reaction with $(CH_3)_3SiCl$ as follows:

$$\equiv SiOH + (CH_3)_3SiCl \rightarrow \equiv SiOSi(CH_3)_3 + HCl$$

In studies of absorption phenomena, it is of interest to know the degree of treatment. This analysis was developed to give such information. Since it is difficult to know the exact amount of powdered sample in the sample beam, an absorbance ratio was used, with the absorbance of the treating material (and the SiOH) compared with a band of the substrate. Calibration for $(CH_3)_3SiO_{1/2}$ was done by carbon analysis. The hydroxyl content of the untreated surface was assumed to be eight micromoles per square meter. The decrease in SiOH absorbance was shown to be directly proportional to the increase in $(CH_3)_3SiO_{1/2}$ absorbance.

Matrix: SiO_2 powder (silica gel or aerosil).

Sample form: powder, dusted between NaCl plates to give 15–40% transmittance at 800 cm^{-1}.

Band selection: $(CH_3)_3SiO_{1/2}$, 2965 cm^{-1} (3.37 μm); SiOH, 940 cm^{-1} (10.6 μm); SiO$_2$ substrate, 800 cm^{-1} (12.5 μm).

Standard preparation: SiO$_2$ was treated with $(CH_3)_3SiCl$, washed, and dried at 110°C overnight. Standards were analyzed for carbon content to obtain a calibration curve.

Calculations: graphical.

Precision: $s = 0.0039$ for $(A_{CH}/A_{Si}) = 0.1131$, or 3.5% relative.

KINETIC STUDIES

The application of infrared techniques to chemical kinetic studies is straightforward, provided that the reaction is not so fast that the response time of the spectrometer limits the accuracy of the measurements. Several methods can be used to make time-dependent studies of concentra-

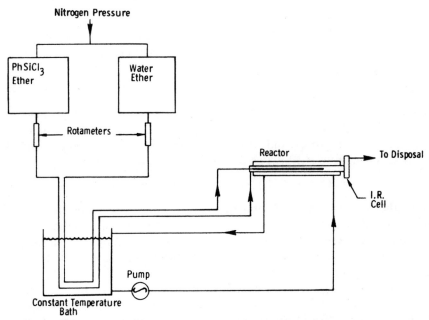

Fig. 6.17 Apparatus for feeding reactants to reactor. Reaction time is varied by changing the flow rate and/or distance of plunger needle from IR cell. Reproduced, with permission, from *Industrial Engineering Chemistry Fundamentals* [23]. Copyright American Chemical Society.

tion. The simplest is to "kill" the reaction in a small aliquot of the reaction mixture by dilution, deactivation of the catalyst or lowering of the temperature, and subsequently scanning the samples. Alternatively, the reaction may be carried out in the IR cell and the spectrum (or a portion of it) scanned repeatedly. If only one constituent is of interest, the spectrometer may be fixed at the frequency of an absorption band and the absorbance observed as a function of time, as illustrated by a study of the thermal decomposition of ethylene oxide [99]. In many cases in which solvents cannot be used, longer sample paths and use of the overtone region may prove satisfactory. Sample temperature, of course, must be closely controlled.

An ingenious cell and mixing system that permits more leisurely scanning of moderately fast reactions (0.1–1 second) with conventional spectrometers is described by Cameron, Kleinhenz, et al. (23) (Figs. 6.17 and 6.18).

Faster scans can be accomplished by interferometer spectrometers. Tracking of the gas-phase nitration of butadiene and NO_2–N_2O_4 by 1-s scans taken at 6.5-s intervals has been reported [69]. A special technique

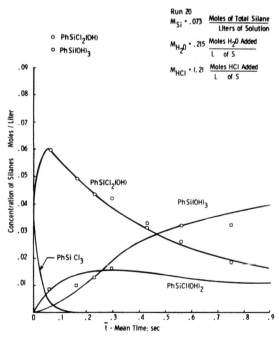

Fig. 6.18 Concentration of silanes versus mean reaction time for hydrolysis of $PhSiCl_3$ [23].

combining spectral and temporal multiplexing on a continuous scan Fourier transform system permits detection of transient species with better than 2-μs resolution [73].

INFRARED PROCESS-STREAM ANALYZERS

It is natural that such a useful laboratory tool as IR spectroscopy should be adopted by the chemical and petroleum industries for continuous analysis of organic mixtures in their manufacturing plants. From analysis it is only a short step to process control, and many physical and chemical operations are directed by the output of an IR monitoring device. Although IR has lost ground to other types of process-stream analyzers, notably GLC, many applications remain for which IR is the best analytical method.

Since the factory environment is usually dirty, corrosive, and subject to vibration, shock, and temperature extremes, a rugged, reliable, enclosed instrument of a special design is used. Instruments covering a wide range of sophistication from simple photometers to multipoint dispersive analyzers have been described. In this section we examine some of the basic designs and discuss briefly their advantages and disadvantages.

Nondispersive Analyzers [35]

The simplest analyzer is the nonselective type, consisting of a source, sample space, radiation detector, and readout, as shown in Fig. 6.19. It is very sensitive, but any IR-absorbing gas will cause a response. The instrument may be made partially selective by use of a filter, either as a separate entity or as a special coating on the receiver. An organic resin, for example, used instead of blackening on the receiver, would sensitize the instrument to absorptions in the C—H stretching region and eliminate

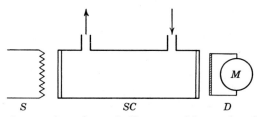

Fig. 6.19 Simple photometric analyzer: S, IR source; SC, sample cell; D, detector; M, readout.

interference from, say, carbon dioxide. The same resin used as a beam filter would have the opposite effect.

This simple photometer is a single-beam instrument and is thus sensitive to source fluctuations, changes in detector sensitivity, and dirt in the sample stream. These effects can be eliminated and the instrument rendered more selective by the addition of another detector and cell as shown in Fig. 6.20. One of the filter cells, F_1, is filled with the gas sought, so radiation passed by this cell is unaffected by concentration changes in the sample cell. The other cell, F_2, may be unfilled or it may contain a suitable pressure of interfering gas, if one is present. The detectors are connected in opposition and the output of this negative filter analyzer is the difference between two large signals. If the two beams can be accurately matched, the analyzer is sensitive to the component sought and is independent of other disturbances. Such an analyzer has been used for analysis of butenes in butadiene gas and of ethyl benzene in liquid styrene monomer [122].

A positive filter concept may be incorporated into an instrument of similar design by the use of a selective receiver. Most commonly, such a detector takes the form of a chamber containing the gas to be analyzed (or a gas with absorption bands coinciding with those of a liquid sample). One wall of the chamber is a thin metal diaphragm that, together with a grid, acts as a condenser microphone [71]. If the radiation is chopped, heating and cooling of the sensitizing gas give pressure changes that induce an alternating signal in the output. This signal is amplified and is used as a measure of the concentration of gas in the sample cell. One such design is shown in Fig. 6.21. Nondispersive gas analyzers have been discussed more completely elsewhere [53]. Systematic procedures for sensitizing nondispersive double-beam analyzers and minimizing interferences have been given [8, 68].

Neither the negative nor the positive filter analyzers are wholly satisfactory if large amounts of interfering absorbers are present, although it is reported that the negative filter instrument allows better compensation for interferences. Both types are more effective when the analytical bands

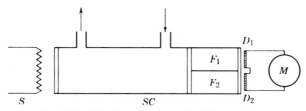

Fig. 6.20 Nondispersive negative filter analyzer.

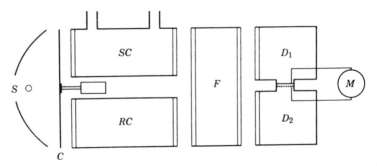

Fig. 6.21 Nondispersive positive filter analyzer: S, source; SC and RC, sample and reference cells, respectively, F, filter cell; D_1 and D_2, detector cells; M, readout.

fall near the peak of the radiation curve, since a larger fraction of the total energy is available to work with. To eliminate some of the shortcomings of the nondispersive analyzers, rugged miniature spectrometers have been developed for in-plant use.

Dielectric filter spectrometers (see pp. 21–22) have also been adapted to process and manufacturing plant service; they are used in the same manner as dispersive analyzers. Stream-switching devices permit sampling at a number of different points, as for example, air monitors in a chemical process building [43].

Dispersive Analyzers

Formidable problems involving analysis of interfering components often may be solved by dispersive analyzers. In fact, almost any analysis that can be done in the laboratory also can be accomplished by such an instrument in an automatic manner. Several types of analyzers, corresponding to single- and double-beam laboratory spectrometers, have been described [11]. Such instruments can be adapted to multicomponent analysis by incorporating a mechanism to sample consecutively the wavelengths of interest. In the case of interferences, a small computer is also incorporated to provide corrected null concentrations.

Double-beam dispersive analyzers offer even greater advantages in selectivity and stability. With such an instrument, interferences can be eliminated effectively by continuous comparison of two wavelengths. The beams may be balanced by means of an optical null device, or electrical ratio recording can be used.

Dispersive analyzers in general are more easily adaptable to a variety of problems and require less adjustment time. Their stability, dependability, and sensitivity are as good as or better than the nondispersive types, and interferences are more easily eliminated. They are more suitable for the limited energy region below 1000 cm^{-1} (10 μm).

The usual problems of sampling and lag time are encountered in IR-analyzer installations. Sample flow through thin liquid cells may be so slow as to require a bypass arrangement through which the main sample stream flows, with a shunt through the IR cell. Or, an ATR crystal may be used, in which case cell thickness is immaterial and a good flow can be maintained through the sample space. In any case, dead time should be kept to a minimum. The sample stream may require filtering to remove dirt, and the analyzer may have to be operated at an elevated temperature to prevent solidification of the sample stream. Almost every sample stream has its own unique problems, and frequently more effort is required to solve sampling difficulties than to adapt the analyzer to the analytical system.

Most analyzers are equipped for periodic automatic or manual standardization using a synthetic sample of known composition. Frequent and careful calibration checks are just as important in process stream analysis as in laboratory analysis. When properly maintained, IR analyzers have a very low percentage of downtime [11].

References

1. Adda, J., E. Blanc-Patin, R. Jeunet, R. Grappin, G. Mocquot, B. Pousardieu, and G. Ricordeau, *Lait,* **48,** 145, 293 (1968).
2. Ahmed, A., and E. Gallei, *Appl. Spectrosc.,* **28,** 430 (1974).
3. American Society for Testing and Materials, *1976 Annual Book of ASTM Standards,* Part 30, ASTM, Philadelphia, Method E 180-67, 1976, p. 538.
4. American Society for Testing and Materials, *1977 Annual Book of ASTM Standards,* Part 42, ASTM, Philadelphia, 1977.
5. Bain, G. H., *Appl. Spectrosc.,* **10,** 193 (1956).
6. Baker, B. B., Jr., *J. Am. Ind. Hyg. Assoc.,* **35,** 735 (1974).
7. Baker, B. B., Jr., and D. P. Carlson, *Anal. Chem.,* **46,** 1583 (1974).
8. Baker, W. J., *Anal. Chem.,* **28,** 1391 (1956).
9. Bard, C. C., T. J. Porro, and H. L. Rees, *Anal. Chem.,* **27,** 12 (1955).
10. Barnett, H. A., and A. Bartoli, *Anal. Chem.,* **32,** 1153 (1960).
11. Bartz, A. M., and H. D. Ruhl, *Chem. Eng. Progr.,* **64**(8), 45 (1968).
12. Basch, A., and E. Tepper, *Appl. Spectrosc.* **27,** 268 (1973).
13. Bauman, R. P., *Absorption Spectroscopy,* Wiley, New York, 1962.
14. Beer, A., *Ann. Physik. Chem.* [2], **86,** 78 (1852).
15. Bentley, F. F., and G. Rappaport, *Charles J. Cleary Awards for Papers on Materials Sciences,* Library of Congress, Washington, 1962, pp. 57–70.

16. Blakeway, J. M., and N. A. Puttnam, *Anal. Chem.*, **35**, 630 (1963).

17. Blout, E. R., M. Parrish, Jr., G. R. Bird, and M. J. Abbate, *J. Opt. Soc. Am.*, **42**, 966 (1952).

18. Bouguer, P., *Essai d'Optique sur la Gradation de la Lumiére,* 1729.

19. Bradley, K. B., and W. J. Potts, Jr., *Appl. Spectrosc.*, **12**, 77 (1958).

20. Brown, R. A., and E. R. Quiram, *Appl. Spectrosc.*, **17**, 33 (1963).

21. Burley, R. A., and W. J. Bennett, *Appl. Spectrosc.*, **14**, 32 (1960).

22. Burns, E. A., and R. F. Muraca, *Anal. Chem.*, **31**, 397 (1959).

23. Cameron, J. H., T. A. Kleinhenz, and M. C. Hawley, *Ind. Eng. Chem., Fund.*, **14**, 328 (1975).

24. Caputi, A., Jr., and M. Udea, *Am. J. Enol. Viticult.*, **24**, 116 (1973).

25. Chaney, C. L., and J. Chin, *Appl. Spectr.*, **28**, 139 (1974).

26. Chaney, L. W., and W. A. McClenny, *Environ. Sci. Technol.*, **11**, 1186 (1977).

27. Cleverley, B., and R. Dolby, *Appl. Spectrosc.*, **32**, 187 (1978).

28. Collier, G. L., and A. C. M. Panting, *Spectrochim. Acta,* **14**, 104 (1959).

29. Critchfield, F. E., and E. T. Bishop, *Anal. Chem.*, **33**, 1034 (1961).

30. Daasch, L. W., *Anal. Chem.*, **19**, 779 (1947).

31. deBruin, H. J., *Anal. Chem.*, **32**, 360 (1960).

32. Diamond, W. J., *Appl. Spectrosc.*, **13**, 77 (1959).

33. Dijkstra, G., *Spectrochim. Acta,* **11**, 618 (1957).

34. Drushcl, H. V., and F. A. Iddings, *Anal. Chem.*, **35**, 28 (1963).

35. Fastie, W. G., and A. H. Pfund, *J. Opt. Soc. Am.*, **37**, 762 (1947).

36. Fochler, H. S., *Appl. Spectrosc.*, **17**, 105 (1963).

37. Forbes, J. W., *Anal. Chem.*, **34**, 1125 (1962).

38. French, A. R., J. V. Benham, and T. J. Pullukat, *Appl. Spectrosc.*, **28**, 477 (1974).

39. Freund, S. M., and D. M. Sweger, *Anal. Chem.*, **47**, 930 (1975).

40. Freund, S. M., D. M. Sweger, and J. C. Travis, *Anal. Chem.*, **48**, 1944 (1976).

41. Friedel, R. A., and J. A. Queiser, *Anal. Chem.*, **29**, 1362 (1957).

42. Fujiyama, T., H. Herrin, and B. L. Crawford, Jr., *Appl. Spectrosc.*, **24**, 9 (1970).

43. Gearhart, H. L., R. L. Cook, R. W. Whitney, and R. D. Hefner, *Anal. Chem.*, **49**, 893 (1977).

44. Golden, B. M., and E. S. Yeung, *Anal. Chem.*, **47**, 2132 (1975).

45. Green, B. D., and J. I. Steinfeld, *Environ. Sci. Technol.*, **10**, 1134 (1976).

46. Hammer, C. F., and H. R. Roe, *Anal. Chem.*, **25**, 668 (1953).

47. Hampton, R. R., *Rub. Chem. Technol.*, **45**, 546 (1972).

48. Hanst, P. L., *Adv. Environ. Sci. Technol.*, **2**, 91 (1971).

49. Hanst, P. L., *Appl. Spectrosc.*, **24**, 161 (1970).

50. Hawranek, J. P., and R. N. Jones, *Spectrochim. Acta* **32A**, 111 (1976).

51. Heigl, J. J., M. F. Bell, and J. U. White, *Anal. Chem.*, **19**, 293 (1947).

52. Herget, W. F., J. A. Jahnke, D. E. Burch, and D. A. Gryvnak, *Appl. Opt.*, **15**, 1222 (1976).

53. Hill, D. W., and T. Powell, *Nondispersive Infrared Gas Analysis in Science, Medicine and Industry*, Hilger, London, 1968.

54. Hilton, C. L., *Anal. Chem.*, **31**, 1610 (1959).

55. Hinkley, E. D., *Environ. Sci. Technol.*, **11**, 564 (1977).

56. Hinkley, E. D., *Opto-electronics*, **4**, 69 (1972).

57. Hirschfeld, T., *Opt. Eng.*, **13**, 15 (1974).

58. Hirschfeld, T., *Res. Devel.*, **27(7)**, 20 (1976).

59. Hollingdale-Smith, P. A., *Can. J. Spectrosc.*, **11**, 107 (1966).

60. Horner, H. J., J. E. Weiler, and N. C. Angelotti, *Anal. Chem.*, **32**, 858 (1960).

61. Huang, A., and D. Firestone, *J. Assoc. Offic. Anal. Chem.*, **54**, 1288 (1971).

62. Hughes, H. K., *Appl. Opt.*, **2**, 937 (1963).

63. Jakeš, J., M. Slabina, and B. Schneider, *Appl. Spectrosc.*, **26**, 389 (1972).

64. Johnson, D. R., J. W. Cassels, E. G. Brame, and D. F. Westneat, *Anal. Chem.*, **34**, 1610 (1962).

65. Jones, R. N., *J. Am. Chem. Soc.*, **74**, 2681 (1952).

66. Jones, R. N., D. Escolar, J. P. Hawranek, P. Neelakantan, and R. P. Young, *J. Molec. Struct.*, **19**, 21 (1973).

67. Kagel, R. O., and J. A. Baker, *Appl. Spectrosc.*, **28**, 65 (1974).

68. Koppius, O. G., *Anal. Chem.*, **23**, 554 (1951).

69. Lephardt, J. O., and G. Vilcins, *Appl. Spectrosc.*, **29**, 221 (1975).

70. Low, M. J. D., and L. Abrams, *Appl. Spectrosc.*, **20**, 416 (1966).

71. Luft, K. F., *Z. Tech. Physik*, **24**, 97 (1943).

72. Malone, C. P. and P. A. Flournoy, *Spectrochim. Acta*, **21**, 1361 (1965).

73. Mantz, A. W., *Appl. Spectrosc.*, **30**, 459 (1976).

74. Mantz, A. W., *Appl. Spectrosc.*, **30**, 539 (1976).

75. Martin, A. E., *Transact. Faraday Soc.*, **47**, 1182 (1951).

76. Maxwell, J. C., C. H. Barlow, J. E. Spallholz, and W. S. Caughey, *Biochem. Biophys. Res. Commun.*, **61**, 230 (1974).

77. McDowell, R. S., *Appl. Spectrosc.*, **26**, 405 (1972).

78. McKelvey, J. M., and H. E. Hoelscher, *Anal. Chem.*, **29**, 123 (1957).

79. Meehan, E. J., Fundamentals of Spectrophotometry, in I. M. Kolthoff and P. J. Elving, Eds., *Treatise on Analytical Chemistry*, Part I, Section D-3, Vol. 5, p. 2753, Interscience, New York, 1964.

80. Murphy, J. F., *Appl. Spectrosc.*, **16**, 139 (1962).

81. Nordstrom, R. J., J. H. Shaw, W. R. Skinner, W. H. Chan, J. G. Calvert, and W. M. Uselman, *Appl. Spectrosc.*, **31**, 224 (1977).

82. Perry, J. A., *Anal. Chem.*, **23**, 495 (1951).

83. Perry, J. A., *Appl. Spectrosc. Rev.*, **3**, 229 (1970).

84. Perry, J. A., and G. H. Bain, *Anal. Chem.*, **29**, 1123 (1957).

85. Perry, J. A., G. H. Bain, and W. B. Traver, *Appl. Spectrosc.*, **10**, 191 (1956).

86. Perry, J. A., R. G. Sutherland, and N. Hadden, *Anal. Chem.*, **22**, 1122 (1950).

87. Pillion, E., M. R. Rogers, and A. M. Kaplan, *Anal. Chem.*, **33**, 1715 (1961).

88. Potts, W. J., Jr., *Chemical Infrared Spectroscopy*, Vol. 1, *Techniques*, Wiley, New York, 1963.

89. Potts, W. J., Jr., and A. L. Smith, *Appl. Opt.*, **6**, 257 (1967).

90. Ramsay, D. A., *J. Am. Chem. Soc.*, **74**, 72 (1952).

91. Robinson, D. Z., *Anal. Chem.*, **23**, 273 (1951).

92. Robinson, D. Z., *Anal. Chem.*, **24**, 619 (1952).

93. Rueda, D. R., F. J. Baltá Calleja, and A. Hidalgo, *Spectrochim. Acta,* **30A**, 1545 (1974).

94. Ruppin, R., and R. Englman, *Rep. Prog. Phys.*, **33**, 149 (1970).

95. Saier, E. L., L. R. Cousins, and M. R. Basila, *Anal. Chem.*, **34**, 824 (1962).

96. Saier, E. L., and R. H. Hughes, *Anal. Chem.*, **30**, 513 (1958).

97. Schnurmann, R., and E. Kendrick, *Anal. Chem.*, **26**, 1263 (1954).

98. Sharpless, N. E., and D. A. Gregory, *Appl. Spectrosc.*, **17**, 47 (1963).

99. Simard, G. L., J. Steger, T. Mariner, D. J. Salley, and V. Z. Williams, *J. Chem. Phys.*, **16**, 836 (1948).

100. Simard, R. G., I. Hasegawa, W. Bandaruk, and C. E. Headington, *Anal. Chem.*, **23**, 1384 (1951).

101. Smith, A. L., *Appl. Spectrosc.*, **12**, 153 (1958).

102. Smith, A. L., Infrared Spectroscopy, in J. W. Robinson, *Handbook of Spectroscopy*, Vol. II, CRC Press, Cleveland, 1974.

103. Smith, A. L., and L. R. Kiley, *Appl. Spectrosc.*, **18**, 38 (1964).

104. Spell, H. L., and R. D. Eddy, *Anal. Chem.*, **32**, 1811 (1960).

105. Sterling, G. B., J. G. Cobler, D. S. Erley, and F. A. Blanchard, *Anal. Chem.*, **31**, 1612 (1959).

106. Stewart, J. E., *Appl. Opt.*, **1**, 75 (1962).

107. Toma, S. Z., and S. A. Goldberg, *Anal. Chem.*, **44**, 431 (1972).

108. Toribara, T. Y., and V. Di Stefano, *Anal. Chem.*, **26**, 1519 (1954).

109. Tosi, C., and F. Ciampelli, *Fortschr. Hochpolym. Forsch.*, **12**, 87 (1973).

110. Tryon, M., and E. Horowitz, Infrared Spectrophotometry, in G. M. Kline, Ed., *Analytical Chemistry of Polymers*, Part II, Interscience, New York, 1962, pp. 291–334.

111. Tuddenham, W. M., and R. J. P. Lyon, *Anal. Chem.*, **32**, 1630 (1960).

112. Walton, W. W., and J. I. Hoffman, Principles and Methods of Sampling, in I. M. Kolthoff and P. J. Elving, Eds., *Treatise on Analytical Chemistry*, Part I, Section A, Vol. 1, p. 67, Interscience, New York, 1959.

113. Westneat, D. F., *Anal. Chem.*, **33**, 812 (1961).

114. Whetsel, K. B., W. E. Roberson, and M. W. Krell, *Anal. Chem.*, **32**, 1281 (1960).

115. White, J. U., and W. M. Ward, *Anal. Chem.*, **37**, 268 (1965).

116. Wiberley, S. E., J. W. Sprague, and J. E. Campbell, *Anal. Chem.*, **29**, 210 (1957).

117. Wiens, R. H., and H. H. Zwick, Trace Gas Detection by Correlation Spectroscopy, in J. S. Mattson, H. B. Mark, Jr., and H. C. MacDonald, Jr., Eds., *Infrared, Correlation, and Fourier Transform Spectroscopy*, Marcel Dekker, New York, 1977.

118. Wilks, P. A., Jr., *Appl. Spectrosc.*, **23**, 63 (1969).

119. Willbourn, A. H., *J. Polym. Sci.*, **34**, 569 (1959).

120. Wood, D. L., and J. P. Luongo, *Mod. Plast.*, **38**, 132 (March, 1961).

121. Wright, N., *Ind. Eng. Chem., Anal. Ed.*, **13**, 1 (1941).

122. Wright, N., and L. W. Hersher, *J. Opt. Soc. Am.*, **36**, 195 (1946).

123. Young, E. F., and R. W. Hannah, in W. W. Wendlandt, Ed., *Modern Aspects of Reflectance Spectroscopy*, Plenum Press, New York, 1968.

124. Zenker, W., *Anal. Chem.*, **44**, 1235 (1972).

CORRELATION CHARTS

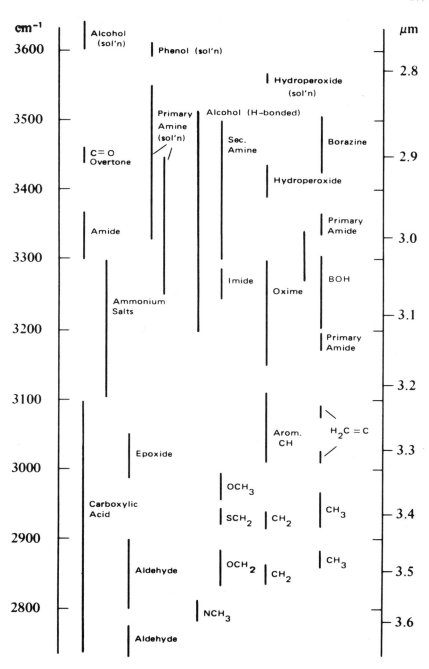

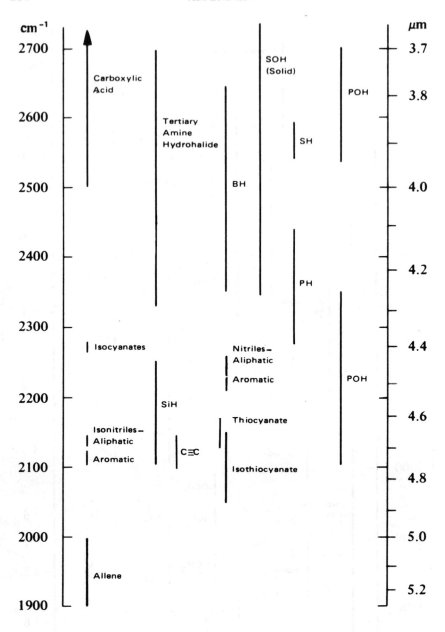

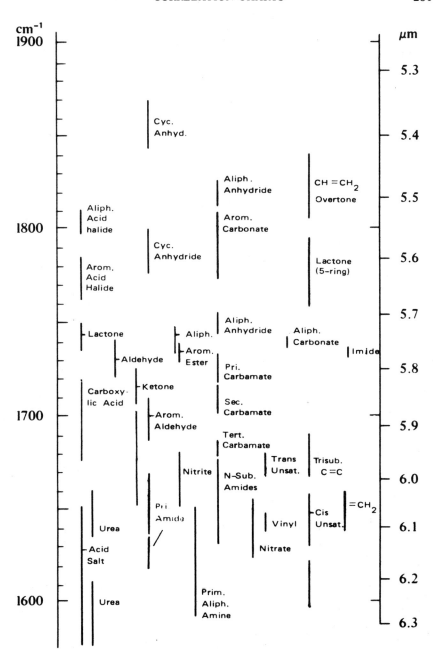

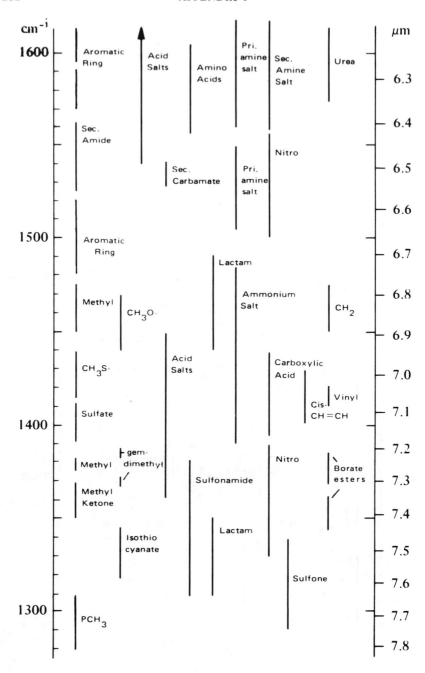

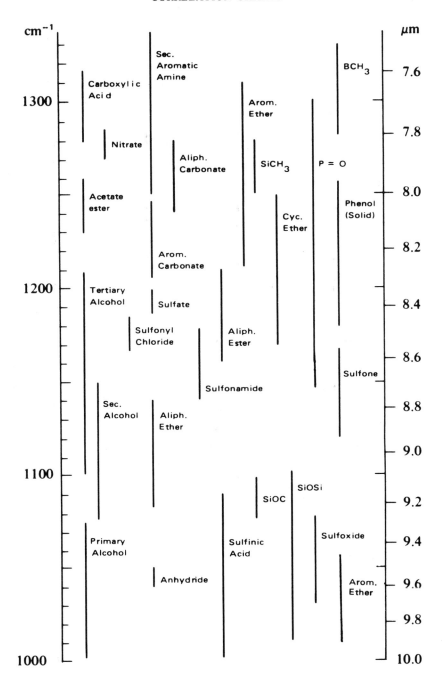

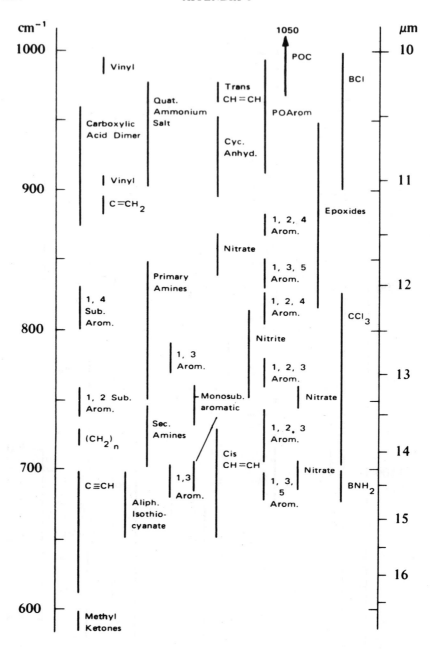

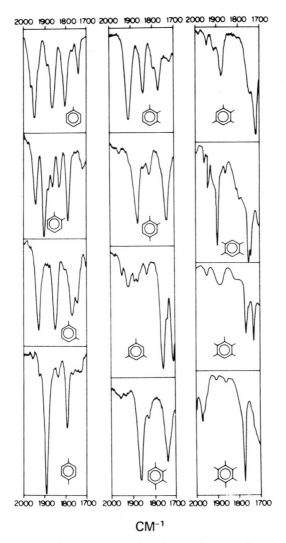

CM⁻¹

Fig. A.21 Characteristic overtone and combination band patterns for substituted benzene derivatives. Bands are weak; intensities shown are observed for ~0.1 mm layer. The *pattern* is more significant than the exact wavenumbers. From C. W. Young, R. B. Du Vall, and N. Wright, *Anal. Chem.*, **23**, 709 (1951). Reproduced with permission of American Chemical Society.

CHARACTERISTIC ABSORPTION FREQUENCIES

Listed here in tabular form are characteristic frequencies for some of the more commonly encountered chemical groups. This listing is not intended to be comprehensive, but only to indicate possible paths for further exploration. Characterization of a pure material by use of its IR spectrum can be accomplished only through the judicious use of reference spectra, and several compilations are referenced in Chapter 3. Monographs giving more complete surveys and discussions of correlations found in the literature are recommended for further reading [1,2] as is the study of model spectra [3].

The following conventions have been used in this listing. All absorption frequencies are given in wavenumbers, usually as a range, for solution in inert solvents unless otherwise specified. Often, the position of the band within the range is related to some structural feature of the molecule. Ranges are approximate; it is likely that a few structures absorb outside the range given. Descriptions of the vibrational modes are also approximate; the COC stretching mode in ethers, for example, involves several other skeletal motions. Some abbreviations are used: ν^a, asymmetric stretching mode; δ^s, symmetrical deformation; ω, wagging vibration; ρ, rocking vibration; R, aliphatic group; Ar, aryl group; X, halogen; A^-, anion; M^+, cation; S, strong (intense band); M, medium; W, weak; V, variable intensity; ip, in-plane; oop, out-of-plane; and comb., combination band.

References

1. Bellamy, L. J., *The Infrared Spectra of Complex Molecules*, Vol. 1, 3rd ed., Wiley, New York, 1975.
2. Colthup, N. B., L. H. Daly, and S. G. Wiberley, *Introduction to Infrared and Raman Spectroscopy*, 2nd ed., Academic Press, New York, 1975.
3. Craver, C. D., *Desk Book of Infrared Spectra*, Coblentz Society, Kirkwood, Mo., 1977.

1. Acids, carboxylic (dimer form)

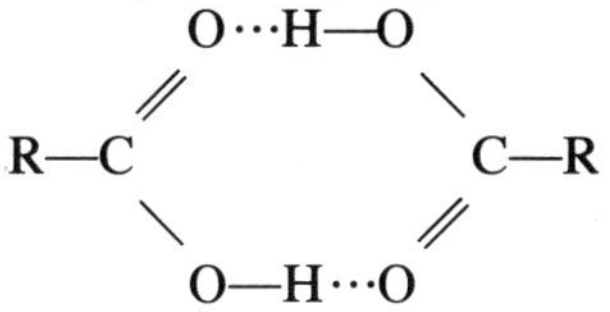

$$\underset{O-H\cdots O}{\overset{O\cdots H-O}{\underset{R-C\diagdown\qquad\diagup C-R}{\diagup\qquad\qquad\diagdown}}}$$

Carboxylic acids are dimerized at room temperature, except in extremely dilute solution.

ν_{OH} 2500–3100 S

Broad. (See Fig. 5.11.) Sharp, weak bands frequently appear near 2500–2700.

$\nu_{C=O}$ 1700–1720 S

Frequency raised by electronegative groups in the α position (15–20 cm^{-1}/Cl or F; 6 cm^{-1} for Br). Conjugation with C=C or aryl group lowers frequency to ~1690.

δ_{OH}(ip) 1395–1440 W

ν_{C-O} 1280–1315 S

δ_{OH}(oop) 875–960 M

Broad.

ω_{CH_2} 1180–1345 W

A progression of bands is seen in the solid state. If N is the number of bands, the number of carbon atoms in the acid is $2N$ (even no.) or $2N - 1$ (odd no. carbons).

Acids, carboxylic (monomer form)

Observed in the vapor state at elevated temperatures, as from a GC effluent, or in very dilute solutions.

ν_{OH} 3500–3580 W

Sharp.

$\nu_{C=O}$ 1740–1800 S

δ_{OH} 1280–1380 M

ν_{C-O} 1075–1190 S

2. Acid halides

$\nu_{C=O}$ 1795–1810 S   Conjugation with double bond or phenyl ring lowers frequency by 25–30 cm^{-1}. Aromatic acid chlorides may have a second weaker band at 1735–1750 (Fermi resonance doublet).

920–965 M

~1200 S   Aryl acid chlorides.

850–890 V

3. Acid salts

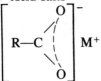

$\nu^{a}_{C=O}$ 1540–1650 S   Frequencies depend on crystal structure and are therefore hard to predict.

$\nu^{s}_{C=O}$ 1360–1450 M

Formates: ~2930, 1600, 1360, 775.

Acetates: 1550–1600, 1400–1450, ~1050, 1020, 925.

Oxalates: $\nu_{C=O}$ ~1620, 1320, 770, 515.

4. Alcohols

  a. Monohydric:

    Aromatic: (see **Phenols**)

    Primary: ν_{OH} ~3640 M   Dilute solution in CCl$_4$.

    RCH$_2$OH   3200–3400 S   Broad. Hydrogen-bonded OH (see Fig. 5.11).

    ν_{CO} 1000–1075 S

[4. Alcohols—continued]

   Secondary: ν_{OH} ~3630 M   CCl₄ solution.

    RR'CHOH   3200–3500 S   Broad.

          ν_{CO} 1075–1150   Straight chain saturated alcohols ~1100; branching lowers frequency by 10–15 cm⁻¹.

   Tertiary: ν_{OH} ~3620 M   CCl₄ solution.

    RR'R"COH   3200–3520 S   Broad.

          ν_{CO} 1100–1210 S   If R = *n*-alkyl, ν_{CO} ~1140; branching, unsaturated or aryl groups lower frequency.

b. Alcohols, polyhydric:

          ν_{OH} ~3600 M   Dilute CCl₄ solution.

          3350–3550 S   Solid or liquid.

c. Carbohydrates:

          ν_{OH} ~3300 S   Broad.

          ν_{C-O} 1000–1125 S   Multiple bands.

5. Aldehydes

   O

   ‖ ν_{CH} 2800–2900 M   ⎫ Fermi resonance doublet (occasionally only one band is seen).

   RCH   2δ_{CH} 2695–2775 M ⎭

          $\nu_{C=O}$ 1720–1740 S   Electronegative substituent on α-carbon raises the frequency (15–20 cm⁻¹ for Cl or F; 6 cm⁻¹ for Br; 10 cm⁻¹ for OR); conjugation lowers it. Hydrogen bonding lowers frequency by 5–10 cm⁻¹.

          δ_{CH} 1380–1410

[5. Aldehydes—continued]

Ar(C=O)H $\nu_{C=O}$ 1685–1710 S

                    1260–1310 W

                    1160–1210 M

6. Amides

Group frequencies are quite variable for amides, since spectra are usually obtained on the solids. Perturbations in the solid state, including hydrogen bonding, are large and often overwhelm other effects.

a. Primary: ν_{NH} ~3350, 3180

In dilute solution, 3520, 3100.

$$\begin{array}{c} O \\ \parallel \\ RCNH_2 \end{array}$$

$\nu_{C=O}$ ~1655 S

δ_{NH_2} ~1630 S

Amide I. ⎫ these bands
Amide II. ⎬ fall at
          ⎭ 1670–1690; 1615–1620.

In dilute solution,

ν_{CN} ~1400 M

$\rho_{NH_2}(ip)$ ~1150 W

(Not always present).

ω_{NH_2} 600–750 W

Broad.

b. N-substituted (trans):

$$\begin{array}{c} O \\ \parallel \\ R-C-N-R' \\ | \\ H \end{array}$$

ν_{NH} ~3300 S

In dilute solution, 3400–3470.

$2\nu_{CN}$ ~3100 W

$\nu_{C=O}$ 1630–1680 S

Amide I. In dilute solution, 1670–1700.

$\nu_{CN}+\delta_{NH}$ ~1550 S

Amide II. Very characteristic. Solution, 1510–1540.

[6. Amides—continued]

| | ν_{CNH} ~1250 V | Amide III. |
| | $\omega_{NH}(oop)$ ~700 W | Broad. |

c. **N-substituted (cis): (see Lactams)**

d. **N-disubstituted:** $\nu_{C=O}$ 1630–1680 S

7. **Amine salts**

a. **Primary:**

$(R—NH_3)^+A^-$	ν_{NH_3} 3200–2800 S	Aromatic amine salts have lower frequency.
	Comb. 2800–2000 W	Usually several bands.
	Comb. ~2000 V	
	$\delta^a_{NH_3}$ 1560–1625 S	
	$\delta^s_{NH_3}$ 1505–1550 S	

b. **Secondary:**

$(R_2NH_2)^+A^-$	ν_{NH_2} 2700–3000 S	Multiple bands.
	Comb. 2300–2700 W	
	δ_{NH_2} 1560–1620 M	

c. **Tertiary:**

| $(R_3NH)^+A^-$ | ν_{NH} 2330–2700 S | Multiple bands. |

d. **Quaternary ammonium compounds:**

| | | No characteristic absorptions, except for —N(CH_3)_3^+. |
| $RN(CH_3)_3^+A^-$ | 900–980 | Multiple bands. Usually also has bands near 3020, 1485, and 1410. |

8. **Amines**

a. **Primary aliphatic** $\nu^a_{NH_2}$ 3330–3350 M $\quad$ $\nu^s = 0.98\nu^a$ for equivalent NH bonds.

RCH_2NH_2 $\quad$ $\nu^s_{NH_2}$ 3250–3450 M $\quad$ Often a shoulder near 3200 also.

[8. Amines—continued]

$\nu^s_{NCH_2}$ 2770–2830 V

δ_{NH_2} 1590–1650 M

ω_{NH_2} 750–850 S   Broad, multiple bands.

b. Primary aromatic

ArNH$_2$ $\nu^a_{NH_2}$ 3420–3520 M   Solution.

$\nu^s_{NH_2}$ 3325–3420 M   Solution.

$\nu^a_{NH_2}$ 3390–3500 M   Liquid.

$\nu^s_{NH_2}$ 3330–3420 M   Liquid. Also weaker band at 3170–3250.

c. Secondary aliphatic

RNHR′ ν_{NH} ~3300 W   Solution.

~3250 W   Liquid.
ω_{NH} 700–750 M

d. Secondary aromatic

ArNHR ν_{NH} ~3400 M

δ_{CHN} ~1510 S   Near the 1500 aromatic band.

ν_{CN} 1250–1340 S

ν_{NR} 1180–1280 M

e. Tertiary amines.   The best way to identify a tertiary amine is to treat it with mineral acid (e.g., 2 drops of sample + 1 drop of concentrated HCl–ethanol) and examine the spectrum of the product in the 2200–3000 region. Absorption near 2600 indicates a tertiary amine hydrochloride (see **Amine salts**).

9. Amino acids (solid)

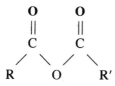

$\nu_{NH_3^+}$ 2600–3100 S

Comb. ~2100 W

$\nu_{CO_2}^a$ 1555–1605 S

$\nu_{CO_2}^s$ 1390–1430 S

Multiple peaks to 2200 cm^{-1}.

10. **Ammonium salts (solid)**
   $(NH_4)^+A^-$

$\nu_{NH_4^+}$ 3100–3330 S

$\delta_{NH_4^+}$ 1390–1485 S

Broad.

11. **Anhydrides**

  a. **Open chain, aliphatic:**

$$\underset{R \quad\quad O \quad\quad R'}{\overset{O \quad\quad\; O}{\underset{\diagup\;\diagdown\;\diagup\;\diagdown}{\overset{\|\quad\quad\|}{\underset{C\quad\quad C}{}}}}}$$

$\nu_{C=O}\begin{cases}\sim1820\ S\\ \sim1750\ M\end{cases}$

ν_{CO} 1040–1050 S

Mechanical coupling causes doubling. Electronegative α substituent raises frequency.

  b. **Open chain, conjugated:**

$\nu_{C=O}\begin{cases}\sim1775\ S\\ \sim1720\ M\end{cases}$

  c. **Cyclic, five-membered ring, nonconjugated:**

$\nu_{C=O}\begin{cases}\sim1860\ W\\ \sim1780\ VS\end{cases}$

ν_{CO} 1000–1130 S

ν_{CO} 895–955 S

  d. **Cyclic, 5-membered ring, conjugated:**

$\nu_{C=O}\begin{cases}\sim1860\ W\\ \sim1760\ S\end{cases}$

ν_{CO} 895–955 S

[11. Anhydrides—continued]

    e. Cyclic, six-membered $\nu_{C=O}$ $\begin{cases} \sim 1800 \text{ M} \\ \sim 1750 \text{ S} \end{cases}$
       ring:

12. Boron
    Compounds

    a. Borate esters: ν_{BO} $\begin{cases} \sim 1380 \text{ S} \\ \sim 1350 \text{ VS} \end{cases}$ Double because of ^{10}B and ^{11}B isotopes.

       $B(OR)_3$ However, no band for

       $RB(OR)_2$ ν_{BO} is seen if N is coordinated to B.

       R_2BOR

    b. BCl
       compounds $\sim 900–1000$ S

    c. Boranes ν_{BH} 2350–2640 S   Normal borane.

       RBH_2 ν_{BHB} 1540–2220 S   Bridged hydrogen in higher borane.

    d. BOH (solid) ν_{OH} 3200–3300 S   Broad.

    e. Borazines: ν_{NH} 3425–3505 M

       R_2BNH_2 ν_{BN} 1330–1470 S

                         δ_{BN} 680–700   Doublet.

    f. BCH_3 $\delta_{CH_3}^{a}$ 1405–1460

                         δ_{CH_3} 1280–1330

    g. BPhenyl   1430–1440 S   Ring vibration.

13. Carbamates

	Condensed	Solution (CHCl₃)

    a. Primary: *Condensed*   *Solution (CHCl₃)*

         O ν_{NH} 3250–3340   3390–3480
         ‖
      H_2NCOR

                     $\nu_{C=O}$ ~ 1705 ~ 1730 if R is aromatic

    b. Secondary: ν_{NH} 3250–3340   3390–3480

      HR'NCOOR $\nu_{C=O}$ 1715–1735   1705–1725
                                1530–1540   1510–1530

[13. Carbamates—continued]

   c. Tertiary:

     R'R''NCOOR $\nu_{C=O}$ 1683–1691

14. Carbonates

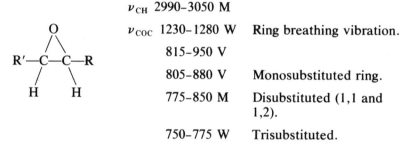

        O
        || $\nu_{C=O}$ ~1800 S
   ArOCOAr ν_{OCO} 1205–1220 S

        O
        || $\nu_{C=O}$ 1720 S
   ArOCOR ν_{OCO} 1210–1250 S

        O
        || $\nu_{C=O}$ ~1740 S
   ROCOR ν_{OCO} 1240–1280 S

15. Epoxides

             ν_{CH} 2990–3050 M

       O ν_{COC} 1230–1280 W   Ring breathing vibration.

      / \\   815–950 V

R'—C—C—R

    /   \\   805–880 V   Monosubstituted ring.

   H   H   775–850 M   Disubstituted (1,1 and 1,2).

                750–775 W   Trisubstituted.

16. Esters

            $2\nu_{C=O}$ 3450 W   Often mistaken for a trace of OH.

     O $\nu_{C=O}$ 1735–1750 S   Conjugation lowers
     || frequency by 10–20 cm^{-1};
   RCOR'   internal H-bonding by 40 cm^{-1}. α-Substituted electronegative group raises frequency, gives double carbonyl band.

            $\nu_{C=O}$ 1720–1725 S   Formate esters

            $\nu_{C=O}$ 1715–1740 S   R=unsaturated or aromatic group.

[16. Esters—continued]

	ν_{CO} 1160–1210 S	Saturated esters except acetates.
	ν_{CO} 1230–1260 S	Acetates only.
	ν_{CO} 1280 S	Aromatic (phthalates, benzoates, etc.).

17. **Ethers**

ROR'

	$\nu^s_{CH_2}$ 2835–2878 V	For aliphatic group next to oxygen.
	ν_{CH_3OAr} ~2835 M	Methoxy group on aromatic ring.
	ν_{COC} 1085–1140 S	Aliphatic ethers. Usually ~1125.
	ν_{COC} 1210–1310 S	Aromatic ethers with alkoxy group.
	ν'_{COC} 1010–1050 M	Aromatic ethers with alkoxy group.
	ν_{COC} 1170–1250 S	Ring compounds (e.g., dioxane).

18. **Fluorocarbons**

   a. **Monofluoro:**

   CF ν_{CF} 1000–1400 M   Not very distinctive.

   b. **Difluoro:**

   CF_2 ν_{CF} 1120–1280 S

   c. **Trifluoro:**

   CF_3 ν_{CF} 1120–1350 S

                680–780 M

19. **Halogenated compounds**

   a. **Chlorine** ν_{CCl} 560–830 W   Not very useful.

                ω_{CH_2Cl} 1240–1300 S

                $\nu^a_{CCl_3}$ 700–830 S

[19. Halogenated compounds—continued]

   b. **Bromine** ν_{CBr} 515–680 M

                ω_{CH_2Br} ~1230 S

   c. **Iodine** ν_{CI} 485–610 M

                ω_{CH_2I} ~1170 S

20. **Heterocyclic compounds**

   a. **Furans, 2-substituted**   1570–1605

                1475–1510

   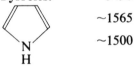—R   1380–1400

                820–885

                740–800

   b. **Pyrroles:**   ~3490   If H-bonded, 3100–3400.

                ~1565

                ~1500

   c. **Thiophenes, 2-substituted**   1514–1534

                1430–1454

                —R   1347–1361

   d. **Thiophenes, 3-substituted**   ~1530

                ~1410

                ~1370

   e. **Pyridines, 2-substituted**   1585–1615

                1568–1576

                1465–1477

                1428–1438

[20. Heterocyclic compounds—continued]

    f. Pyridines, 3-substituted   1590–1600

                                        1572–1582

                                        1465–1485

                                        1417–1425

    g. Pyridines, 4-substituted   1600–1610

                                          1552–1570

                                        1480–1520

                                        1410–1420

    h. Pyridine *N*-oxides   1200–1320

                                        845–880

21. Hydrocarbons

    a. Saturated aliphatic:

Methyl groups

RCH_3	$\nu^a_{CH_3}$ 2952–2972 S	
	$\nu^s_{CH_3}$ 2862–2882 M	
	$\delta^a_{CH_3}$ 1450–1475 M	
	$\delta^s_{CH_3}$ 1377–1383 M	
$ArCH_3$	$\nu^a_{CH_3}$ 2920–2930 M	
	$\nu^s_{CH_3}$ 2860–2870 M	
$RCH(CH_3)_2$	δ^s $\begin{cases} \sim1385 \text{ M} \\ \sim1370 \text{ M} \end{cases}$	*gem*-Dimethyl doublet
	~1170 M	
	1130–1150 V	
$RCH(CH_3)_2R'$	1151–1159 M	
$RC(CH_3)_2R'$	δ^s $\begin{cases} \sim1215 \text{ M} \\ \sim1195 \text{ M} \end{cases}$	

[21. Hydrocarbons—continued]

$RC(CH_3)_3$ $\delta^s \begin{cases} \sim 1400 \text{ M} \\ \sim 1370 \text{ MS} \end{cases}$

1235–1255 M

1160–1210 M

O
‖
$RCCH_3$ ν_{CH_3} 2900–3000 W

δ_{CH_3} 1405–1440 M

δ_{CH_3} 1350–1375 S

$RNCH_3$ ν 2780–2805 S

$ArNCH_3$ ν 2810–2820

$ROCH_3$ ν^a 2955–2992 S

ν^s 2865–2897 S

2δ 2815–2840 V

δ 1440–1470 M

$RSCH_3$ δ^a 1415–1440 M

δ^s 1290–1330 W

Methylene groups

RCH_2R' $\nu^a_{CH_2}$ 2916–2936 S

$\nu^s_{CH_2}$ 2843–2863 M

$\delta^s_{CH_2}$ 1450–1475 M

$R(CH_2)_{n>3}$ ρ_{CH_2} 772–730 W   May be double in solids.

O
‖
$\left.\begin{matrix} RCH_2CR' \\ RCH_2C{\equiv}N \\ RCH_2NO_2 \end{matrix}\right\}$ $\begin{cases} \nu \text{ 2900–3000 M} \\ \delta \text{ 1405–1445 S} \end{cases}$

$\left.\begin{matrix} RCH_2OR' \\ RCH_2OH \\ RCH_2NH_2 \end{matrix}\right\}$ $\begin{cases} \nu^a \text{ 2922–2955 S} \\ \nu^s \text{ 2835–2880 S} \\ \delta \text{ 1445–1475 M} \end{cases}$

[21. Hydrocarbons—continued]

$\left.\begin{array}{l} RCH_2NHR' \\ RCH_2NR'_2 \end{array}\right\}$ $\left\{\begin{array}{l} \nu^a\ 2920-2960\ S \\ \nu^s\ 2760-2820\ S \\ \delta\ 1445-1475\ M \end{array}\right.$

RCH_2SR' $\delta\ 1415-1440\ M$

               $\omega\ 1220-1270\ S$

RCH_2Cl $\delta\ 1430-1460\ M$

               $\omega\ 1240-1300\ S$

Isolated CH

R_3CH $\left.\begin{array}{l} 2890\ W \\ 1340\ W \end{array}\right\}$ Of no practical use.

b. Alkene: $\nu^a\ 3077-3092\ M$

Vinyl: $\nu^s\ 3012-3025\ M$

$RCH{=}CH_2$ $2\omega\ 1805-1840\ W$

               $\nu_{C=C}\ 1638-1648\ M$

               $\delta_{CH_2}\ 1412-1420\ M$

               $\omega\ 985-995\ S$ Varies with inductive effect of R group (see Fig. 5.8).

               $\omega\ 905-910\ S$ Varies with resonance properties of substituents.

trans $\nu\ 3000-3050\ M$

$RCH{=}CHR'$ $\nu_{C=C}\ 1668-1678\ W$

               $\omega\ 965-980\ S$ Varies with sum of substituent inductive effects. Drops by 40 cm^{-1} for each Cl, O, etc.

cis $\nu_{C=C}\ 1630-1660\ M$

$RCH{=}CHR'$ $\rho_{CH}\ 1440-1430\ M$

               $\omega\ 650-730\ M{-}S$

[21. Hydrocarbons—continued]

Vinylidine:

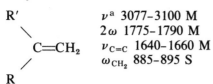

ν^a 3077–3100 M	
2ω 1775–1790 M	Negative anharmonicity.
$\nu_{C=C}$ 1640–1660 M	
ω_{CH_2} 885–895 S	Varies with resonance properties of substituents.

Trisubstituted:

RR′C=CHR″ ν 2990–3050 W

$\nu_{C\ C}$ 1667–1692 W

ω 790–840 M

c. Alkyne

RC≡CH ν_{CH} 3270–3340 S

$\nu_{C\equiv C}$ 2100–2140 W

δ_{CH} 610–700 M

d. Allene:

RC=C=CR′ $\nu_{C=C}$ 1940–1950   May be doublet if R is electron-withdrawing group.

δ_{CH_2} ~850

e. Aromatic:

ν_{CH} 3010–3110 M	Three or four peaks.
1660–2000 W	Characteristic overtone patterns (see Fig. A.2.1).
ν_{CC} ~1600 V ~1580 V	Ring modes. Forbidden if molecule has center of symmetry. Weak for nonpolar substituents, stronger for polar-substituted; 1580 band especially strong for O- or N-substituted rings.

[21. Hydrocarbons—continued]

ν_{CC}	~1490 V ⎫ ~1450 V ⎭	Ring modes. Intensity 1490 strong for electron-donor groups. In *para*-substituted molecules, these bands fall near 1500 and 1370.
δ_{CH}	730–885	Frequencies more characteristic of the position of the substituents than of their nature. However, the correlations become unreliable for strongly polar substituents such as NO_2 or COOH.

Monosubstituted	735–765; 685–710
1,2-disubstituted (*ortho*)	740–760
1,3-disubstituted (*meta*)	770–790; 680–705
1,4-disubstituted (*para*)	800–830
1,2,3 trisubstituted	760–780; 705–745
1,2,4 trisubstituted	870–885; 805–825
1,3,5 trisubstituted	830–850; 680–700
1,2,3,4 tetrasubstituted	~820
1,2,3,5 tetrasubstituted	~865
1,2,4,5 tetrasubstituted	~865
1,2,3,4,5 pentasubstituted	~865

22. Hydroperoxides

ROOH	ν_{OH}	~3560	CCl_4 solution (free OH).
		3390–3435	Hydrogen-bonded.
		~815	Two bands if the alkyl group has an even number of carbons.
RR′R″COOH		830–845	

23. Imides

ν_{NH} ~3280	Solid state. Noncyclic.
$\nu_{C=O}$ 1733–1737 S	Solid state. Noncyclic.
1503–1505	
1167–1236	
732–739	

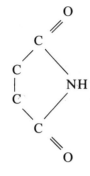

$\nu_{C=O}$ 1735–1790 M

1680–1745 S

24. Isocyanates

R-NCO ν^a 2263–2275 S

25. Isocyanurates

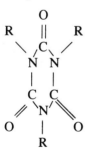

$\nu_{C=O}$ 1680–1700 S   Also shoulder near 1755.

$\nu_{C=O}$ 1710–1715 S   Aromatic substituted.
Shoulder ~1780.

26. Isonitriles (isocyanides)

RNC ν 2134–2146 S

ArNC ν 2110–2130 S

27. Isothiocyanates

RNCS ν^a 2050–2150 S

           ν^s 650–700

[27. Isothiocyanates—continued]

RCH$_2$NCS δ_{CH_2} 1318–1347 S

ArNCS ν^s 925–945

28. Ketones

O $\nu_{C=O}$ 1705–1725 S   Substituents on α-carbon
‖   raise frequency ~15–20
RCR′   cm^{-1}/F or Cl; ~6 cm^{-1}/Br;
                                        ~10 cm^{-1}/OR. Doubling
                                        may result from rotational
                                        isomers. Conjugation (one
                                        group) lowers frequency
                                        by 30–40 cm^{-1}; two
                                        groups by ~60 cm^{-1}. H-
                                        bonding drops frequency
                                        by 5–10 cm^{-1}.

O δ_{CH_3} 1350–1370 S
‖ $\delta_{C=O}$ ~595
RCCH$_3$

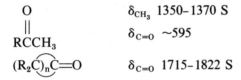 (R$_2$C)$_n$C=O $\delta_{C=O}$ 1715–1822 S   Frequency depends on
                                        ring size because of
                                        mechanical interaction
                                        with CC stretches.

29. Lactams (*cis*-N-Substituted amides)

N—H ν_{NH} ~3200 S
| comb. ~3100 M
C=O

                $\nu_{C=O}$ ~1650   For six- to seven-
                                        membered rings.
                1700–1560   For five-membered rings.
                1730–1760   For four-membered rings.
                δ_{NH} 1440–1490
                ν_{CN} 1310–1350
                ω_{NH} ~800   Broad.

30. Lactones

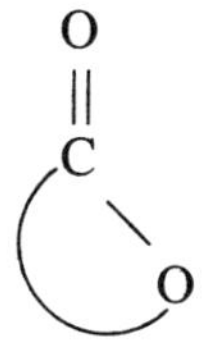

O
‖
C
(ring with O)

$\nu_{C=O}$ 1735–1750 S Six-membered rings.

1760–1795 S Five-membered rings. If unsaturation is present in the ring, two bands are seen in the carbonyl region.

31. Mercaptans

RSH

ν_{SH} 2540–2590 W

ν_{CS} 550–700 W

32. Nitrates (covalent)

$RONO_2$

$\nu^a_{NO_2}$ 1625–1660 S

$\nu^s_{NO_2}$ 1270–1285 S

ν_{NO} 840–870 S

δ 745–760 S

δ 690–710 S

33. Nitriles

R—C≡N

ν_{CN} 2240–2260 M Lowered by conjugation to 2215–2235.

ArC≡N

ν_{CN} 2220–2240 V Frequency and intensity depend on aromatic ring substituents.

34. Nitrites (covalent)

R—O—N=O

ν_{NO} 1648–1681 S (*trans*)

1605–1625 M (*cis*)

ν_{NO} 750–814 S

35. Nitro Compounds

RCH_2NO_2

$\nu^a_{NO_2}$ 1545–1556 S Electronegative substituents affect the frequency [see equation (5.31)].

$\nu^s_{NO_2}$ 1368–1390 M

[35. Nitro compounds—continued]

$RCHXNO_2$ ν^a 1556–1580 S

                  ν^s 1340–1368 M

$ArNO_2$ ν^a 1500–1530 S

                  ν^s 1330–1370 M

36. Nitrosamines

R_2N—N=O ν_{NO} 1425–1460

                  ν_{NN} 1030–1150

$ArRN$—N=O ν_{NO} 1450–1500

                  ν_{CN} 1160–1200

                  ν_{NN} 925–1025

37. Nitroso Compounds

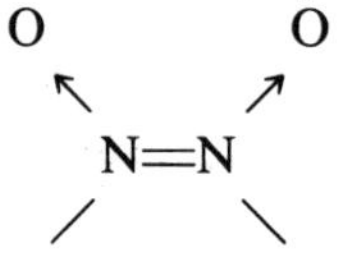

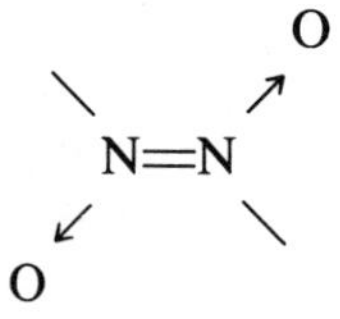

ν_{NO} 1330–1420 S	Aliphatic. Usually found as dimers.	
ν_{NO} 1323–1344 S	Aliphatic, in solid state.	
ν 1390–1397 S	Aromatic.	
ν ~1409 S	Aromatic.	
ν 1176–1290 S	Aliphatic.	
ν 1253–1300 S	Aliphatic.	

38. Oximes

R_2—C=N—OH ν_{OH} 3150–3300

                  ν_{CN} 1620–1690

                  ν_{NO} ~930

Peroxides

RCH_2OOCH_2R' ν_{OO} 830–1000 W   Of no use in identification.

39. Phenols

ArOH	ν_{OH} 3593–3617 M	Dilute solution (free OH).
	ν_{OH} 3200–3250 V	(Bonded OH).
	δ_{OH} 1330–1390 M	Solid State. Frequencies lower in solution.
	ν_{CO} 1180–1260 S	Solid State.

40. Phosphorus Compounds

Many bands occur as doublets because of rotational isomerism in phosphorus compounds.

a. Phosphines:

R_nPH_{3-n}	ν_{PH} 2275–2440 M	Sharp band.
	δ_{PH_2} 1080–1090 M	
	ω_{PH_2} 910–940	
	δ_{CH_2P} 1405–1440	

b. Phosphine oxides:

$R_3P{\rightarrow}O$	ν_{PO} 1140–1300 S	Frequency varies with the sum of the substituent electronegativities.
	ν_{PO} ~1150 S	Aliphatic.
	~1190 S	Aromatic.

c. Phosphorus acids:

$R_n(HO)_{3-n}PO$	ν_{OH} 2550–2700 M	Broad.
	2100–2350 M	

d. Phosphate esters:

$(RO)_3P{\rightarrow}O$	ν_{PO} ~1280 S	
	ν_{COP} 970–1050 S	
	740–830 S	Methoxy and ethoxy derivatives only.

[**40. Phosphorus compounds—continued**]

$(ArO)_3P{\rightarrow}O$ ν_{CO} 1160–1260 S

   ν_{PO} 914–994 S

e. Phosphites:

$(RO)_3P$ ν_{PO} 855–875

f. Other Correlations

P^VF	815–890
PCl	440–580
PBr	400–485
POP	870–1000
$PO_3^=$	970–1030
$PO_4^{\equiv}$	1000–1100
$POCH_3$	~2960; 1460; 1170–1190
POC_2H_5	~2990; 1485; 1450; 1395; 1375; 1155–1165
POC_6H_5	1160–1240; 855–994
PCH_2	1405–1440
PCH_3	1280–1310; 860–960
$P{\rightarrow}S$	580–750
P—N	930–1110
P≡N(cyclic)	1100–1320

41. Silicon Organic Compounds

a. Silanes:

R_nSiH_{4-n} ν_{SiH} 2100–2250 S   Exact position depends on inductive effect of substituents.

   δ_{SiH} 800–840 S

   δ_{SiH_2} 840–900 S

   920–940 S

[41. Silicon organic compounds—continued]

$$\delta_{SiH_3} \quad 910\text{–}930 \text{ S}$$
$$930\text{–}945 \text{ S}$$

b. **Siloxanes:** ν_{SiOSi} 1000–1100 S   Long chain polymers have two strong, broad bands at ~1080 and 1020.

cyclo-$(R_2SiO)_3$ ν_{SiOSi} ~1020 S

cyclo-$(R_2SiO)_4$ ν_{SiOSi} ~1080 S

cyclo-$(R_2SiO)_5$ ν_{SiOSi} ~1080 S

c. **Other correlations:**

$SiCH_3$ δ_{CH_3} 1250–1280 S

760–860 S

SiC_2H_5 1220–1250 M

1000–1020 M

945–970 M

SiC_6H_5 1430 M

1100–1125 S

~730 M

690–700 M

$SiOCH_3$ 2840 S   Hydrolyzable group.

1190 M

1090 S

800–850 M

SiOR   1075–1100 S

945–990 M

SiOAr   920–970 M

SiOH ν_{OH} ~3960 W   Dilute solution.

3200–3400 M   Hydrogen bonded.

δ_{OH} 830–950 M

$SiCH{=}CH_2$ ν_{CC} 1590–1610 M

~1400 M

[**41. Silicon organic compounds—continued**]

	990–1020 M	
	940–980 M	
SiCH$_2$Si	1040–1080 S	
SiNHSi	ν_{NH} 3390–3420 M	Hydrolyzable group.
	~1170 S	
	910–950 S	
SiNH$_2$	ν_{NH} 3480;3400 M	Hydrolyzable group.
	1540 M	
SiOCOCH$_3$	$\nu_{C=O}$ 1700–1770 S	
	1190–1250 S	
	1000–1050 S	
	925–970 S	
SiCl	ν 460–650 S	Hydrolyzable group.
SiCl$_2$	ν^a 535–600 S	
	ν^s 460–540 M	
SiCl$_3$	ν^a 570–620 S	
	ν^s 450–535 M	

42. Sulfates

$$\begin{array}{c} O \\ \| \\ ROSOR' \\ \| \\ O \end{array}$$

ν^a 1390–1415 S

ν^s 1187–1200 S

43. Sulfides

RSR'	ν 570–705 W	Of little use; Raman spectroscopy a better technique.
RSSR'	ν ~500 W	Disulfide. Of little use.
RCH$_2$S—	$\nu^a_{CH_2}$ 2922–2948	
	ν^s 2846–2878	

[**43. Sulfides—continued**]

δ 1410–1435

ω 1220–1270 S

CH$_3$S— δ^a 1415–1440

δ^s 1290–1330

ρ 960–1030

ArS—   ~1090

44. Sulfinic acids

O
‖
RSOH

ν_{OH} 2340–2790   Solid state.

$\nu_{S=O}$ 990–1090

ν_{SO} 810–870

45. Sulfites

O
‖
ROSOR'

$\nu_{S=O}$ 1198–1220

46. Sulfonamides

O
‖ ╱
RS N
‖ ╲
O

ν_{SO}^a 1310–1380 S

ν_{SO}^s 1140–1180 S

ν_{NH_2} 3300, 3250   Primary sulfonamides.

δ ~1570

ν_{SN} 900–910

ν_{NH} 3270–3330   Secondary sulfonamides.

47. Sulfones

O
‖
RSR'
‖
O

ν_{SO}^a 1290–1340 S   Frequently a triplet in
ν_{SO}^s 1120–1165 S   CCl$_4$ solution.

δ_{SO_2} 500 610 M

ω_{SO_2} 460–525 M

48. Sulfonic Acids

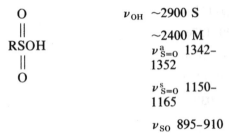

ν_{OH} ~2900 S

~2400 M

$\nu^a_{S=O}$ 1342–1352

$\nu^s_{S=O}$ 1150–1165

ν_{SO} 895–910

Broad. These compounds hydrate very easily to form hydronium sulfates, whose spectra resemble those of the sulfonate salts.

49. Sulfonic Acid Salts

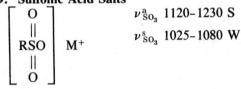

$\nu^a_{SO_3}$ 1120–1230 S

$\nu^s_{SO_3}$ 1025–1080 W

$(ArSO_3)^-M^+$ ~1230,1190,1130,1040

50. Sulfonyl Chlorides

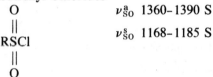

ν^a_{SO} 1360–1390 S

ν^s_{SO} 1168–1185 S

51. Sulfoxides

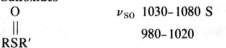

ν_{SO} 1030–1080 S

980–1020   If hydrogen bonded.

52. Thiocarbonyl Compounds

$$\underset{\overset{\|}{-C-}}{\overset{S}{}}$$

$\nu_{C=S}$ 700–1420 V

Not as useful as C=O band. Stretching vibration frequently mixed with other modes.

a. Trithiocarbonates:

$$\underset{RSCSR'}{\overset{S}{\overset{\|}{}}}$$

$\nu_{C=S}$ ~1070

[52. Thiocarbonyl compounds—continued]

   b. **Dithioesters:**

$$\begin{matrix} S \\ \| \\ RCSR' \end{matrix}$$ 1170–1225

   c. **Xanthates:**

$$\begin{matrix} S \\ \| \\ ROCSX \end{matrix}$$ 1200–1250 S

1110–1140 M

1020–1070 S

   d. **Thioamides:**

$$\begin{matrix} S \\ \| \\ RCN \end{matrix}$$ ν_{NH_2} ~3380   Solid state.

~3180

δ ~1630

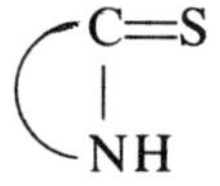

$\begin{matrix} C{=}S \\ | \\ NH \end{matrix}$ ~1550

53. **Thiocyanates**

   R—SCN ν_{SCN} 2160–2175 M

54. **Thiolesters**

$$\begin{matrix} O \\ \| \\ RCSR' \end{matrix}$$ $\nu_{C=O}$ 1680–1700

ν_{CS} 930–1030

$$\begin{matrix} O \\ \| \\ RCSAr \end{matrix}$$ 1690–1710

920–1020

$$\begin{matrix} O \\ \| \\ ArCSR \end{matrix}$$ 1640–1680

940–905

$$\begin{matrix} O \\ \| \\ ArCSAr' \end{matrix}$$ 1650–1700

895–920

$$\begin{matrix} O \\ \| \\ HCSR \end{matrix}$$ ~1675

~755

55. Urea Derivatives

$$\begin{matrix} & O \\ & \| \\ RNC\,NR' \\ H\ \ H \end{matrix}$$

$\nu_{NH}\ \sim\!3300$

$\nu_{C=O}\ 1635\!-\!1660$

$1575\!-\!1613$

Ureas show spectra
similar to those of amides.

Index

315